Smart Technologies for Improved Performance of Manufacturing Systems and Services

This book discusses smart technologies and their influence in the field of manufacturing and industrial systems engineering, in the context of performability enhancement, and explores the development of the workforce for the execution of such smart and advanced technologies.

Smart Technologies for Improved Performance of Manufacturing Systems and Services discusses the integration of smart technology into the production process and supply chain to enhance the overall performance of manufacturing industries. As well as emphasizing the fundamentals of smart technologies, such as artificial intelligence, big data, and cyber-physical systems, it highlights the role that machine learning plays along with other smart technologies. Real-time case studies highlight the applications of smart digital technologies, while research insights into the area of performability and overall sustainable development round out the great range of discussions this reference book has to offer.

Managers and stakeholders seeking coverage on techniques and methods for integration into their organizations, as well as students and researchers in the field, will find this book very useful.

Advances in Intelligent Decision-Making, Systems Engineering, and Project Management

This new book series will report the latest research and developments in the field of information technology, engineering and manufacturing, construction, consulting, healthcare, military applications, production, networks, traffic management, crisis response, human interfaces, and other related and applied fields. It will cover all project types, such as organizational development, strategy, product development, engineer-to-order manufacturing, infrastructure and systems delivery, and industries and industry-sectors where projects take place, such as professional services, and the public sector including international development and cooperation, etc. This new series will publish research on all fields of information technology, engineering, and manufacturing including the growth and testing of new computational methods, the management and analysis of different types of data, and the implementation of novel engineering applications in all areas of information technology and engineering. It will also publish on inventive treatment methodologies, diagnosis tools and techniques, and the best practices for managers, practitioners, and consultants in a wide range of organizations and fields including police, defense, procurement, communications, transport, management, electrical, electronic, aerospace, requirements.

Blockchain Technology for Data Privacy Management
Edited by Sudhir Kumar Sharma, Bharat Bhushan, Aditya Khamparia, Parma Nand Astya, and Narayan C. Debnath

Smart Sensor Networks Using AI for Industry 4.0
Applications and New Opportunities
Edited by Soumya Ranjan Nayak, Biswa Mohan Sahoo, Muthukumaran Malarvel, and Jibitesh Mishra

Hybrid Intelligence for Smart Grid Systems
Edited by Seelam VSV Prabhu Deva Kumar, Shyam Akashe, Hee-Je Kim, and Chinmay Chakrabarty

Machine Learning-Based Fault Diagnosis for Industrial Engineering Systems
Rui Yang and Maiying Zhong

Smart Technologies for Improved Performance of Manufacturing Systems and Services
Edited by Bikash Chandra Behera, Bikash Ranjan Moharana, Kamalakanta Muduli, and Sardar M. N. Islam

For more information about this series, please visit: https://www.routledge.com/ Advances-in-Intelligent-Decision-Making-Systems-Engineering-and-Project-Management/book-series/CRCAIDMSEPM

Smart Technologies for Improved Performance of Manufacturing Systems and Services

Edited by
Bikash Chandra Behera
Bikash Ranjan Moharana
Kamalakanta Muduli
Sardar M. N. Islam

CRC Press
Taylor & Francis Group
Boca Raton London New York

CRC Press is an imprint of the
Taylor & Francis Group, an **informa** business

Cover image: Shutterstock ©

First edition published 2024
by CRC Press
2385 Executive Center Drive, Suite 320, Boca Raton, FL 33431

and by CRC Press
4 Park Square, Milton Park, Abingdon, Oxon, OX14 4RN

CRC Press is an imprint of Taylor & Francis Group, LLC

ISBN: 978-1-032-38689-8 (hbk)
ISBN: 978-1-032-38759-8 (pbk)
ISBN: 978-1-003-34662-3 (ebk)

DOI: 10.1201/9781003346623

Typeset in Times
by SPi Technologies India Pvt Ltd (Straive)

Contents

Editors

Dr. Bikash Chandra Behera is an Assistant Professor at the Department of Mechanical Engineering, C.V. Raman Global University, India. He completed his Ph.D. in Mechanical Engineering with distinction from the Indian Institute of Technology Delhi, India. He is currently working in the area of Artificial Intelligence and Machine Learning applications in manufacturing processes. He is known for his extensive contributions to the field of machining and has made significant advancements in eco-friendly machining techniques. His contributions to the field of machining have resulted in patents, books, book chapters, and numerous research papers published in prestigious international journals and conference proceedings.

Dr. Bikash Ranjan Moharana is an Assistant Professor at the Mechanical Engineering Department, at C.V. Raman Global University, India. He received his Ph.D. degree at the Department of Mechanical Engineering, National Institute of Technology, India. His research interests include various fusion welding processes, non-traditional machining, process optimization, and mechanical and metallurgical analysis. He is a fellow member in Institution of Engineers India, IIW and IWS.

Dr. Kamalakanta Muduli, is an Associate Professor at the Department of Mechanical Engineering, Papua New Guinea University of Technology, Papua New Guinea. He obtained his Ph.D. from the School of Mechanical Sciences, Indian Institute of Technology Bhubaneswar, India. He is a recipient of the ERASMUS+ KA107 award granted by the European Union. His research interests include materials science, manufacturing, sustainable supply chain management, and industry 4.0 applications in operations and supply chain management. Dr Muduli is a fellow of the Institution of Engineers India and a senior member of the Indian Institution of Industrial Engineering and member of ASME.

Professor Sardar M.N. Islam, Ph.D., LL.B. (Law), is a Professor at Victoria University, Australia. He has published extensively across a broad range of disciplines, and his research has attracted international acclaim, leading to a large number of appointments as a distinguished visiting professor, visiting professor, or adjunct professor in different countries, as well as a keynote speaker at many international conferences.

Contributors

G. K. Agrawal
Department of Mechanical Engineering
GEC
Bilaspur, India

Bikash Chandra Behera
Department of Mechanical
Engineering
C.V. Raman Global University
Bhubaneswar, India

Anmol Bhengra
School of Mechanical Engineering
KIIT Deemed University
Bhubaneswar, India

Dillip Kumar Biswal
Krupajal Engineering College
Bhubaneswar, Odisha, India

Pravatnallini Chhotaray
Department of Mechanical
Engineering
C.V. Raman Global University
Bhubaneswar, India

Jayashree Das
Mechanical Engineering Department
Indian Institute of Technology
Guwahati, India

Soumyajit Das
School of Mechanical Engineering
KIIT Deemed University
Bhubaneswar, India

Shailesh Dewangan
Department of Mechanical Engineering
Chouksey Engineering College Bilaspur
C. G., India

Granville James Embia
Mechanical Engineering Department
Papua New Guinea University of
Technology
Lae, Papua New Guinea

Kialakun N. Galgal
Department of Mechanical Engineering
Papua New Guinea University of
Technology
Lae, Papua New Guinea

Manoj Kumar Gopaliya
Department of Mechanical Engineering
The NorthCap University
Gurgaon, Haryana, India

Shiv Manjaree Gopaliya
School of Mechanical Engineering
VIT Bhopal University
Sehore, India

Trupti R. Mahapatra
Department of Production Engineering
Veer Surendra University of Technology
Odisha, India

Raju Prasad Mahto
Sardar Vallabhbhai National Institute of
Technology
Surat, India

Debadutta Mishra
Department of Production Engineering
Veer Surendra University of Technology
Odisha, India

Bikash Ranjan Moharana
Department of Mechanical Engineering,
C.V. Raman Global University
Bhubaneswar, India

Noor Hafizah Mohmaed
Department of Mechanical
    Engineering
Universiti Malaysia
Perlis, Malaysia

Kamalakanta Muduli
Department of Mechanical
    Engineering
Papua New Guinea University of
    Technology
Lae, Papua New Guinea

Suryakant Muduli
Department of Production
    Engineering
Veer Surendra University of
    Technology
Odisha, India

Allu V. K. Murty
AIML Architect
Bengaluru, India

Shubhrata Nagpal
Department of Mechanical Engineering
BIT Durg
Chhattisgarh, India

Chinmaya Prasad Nanda
School of Mechanical Engineering
KIIT Deemed University
Bhubaneswar, India

Pratap Chandra Padhi
CIPET: Institute of Petrochemicals
    Technology (IPT)
Bhubaneswar, Odisha, India

Soumya R. Parimanik
Department of Production Engineering
Veer Surendra University of
    Technology
Odisha, India

Deepak Patil
Department of Production Engineering
National Institute of Technology
    Tiruchirappalli
Tiruchirappalli, India

Ashutosh Pattanaik
Department of Mechanical Engineering
JAIN Deemed to be University
Bengaluru, India

Manidatta Ray
Decision Science, Operations
    Management and IT Area
Birla Global University
Bhubaneswar, India

Bharat Chandra Routara
School of Mechanical Engineering
KIIT Deemed to be University
Bhubaneswar, India

Matruprasad Rout
Department of Production Engineering
National Institute of Technology
    Tiruchirappalli
Tiruchirappalli, India

B. N. Sahoo
Sardar Vallabhbhai National Institute of
    Technology
Surat, India

Prakash Kumar Sahu
Production Engineering Department
National Institute of Technology
    Agartala
Tripura, India

Mantra Prasad Satpathy
School of Mechanical Engineering
KIIT Deemed University
Bhubaneswar, India

Fono-Tamo Romeo Sephyrin
Tickle College of Engineering
University of Tennessee
Knoxville, Tennessee

Qingyu Shi
Department of Mechanical Engineering
Tsinghua University
Beijing, China

Sharda Pratap Shrivas
Department of Mechanical Engineering
BIT Durg
Chhattisgarh, India

Harsh Soni
Sardar Vallabhbhai National Institute of
   Technology
Surat, India

Abbreviations

AI	Artificial Intelligence
AM	Additive Manufacturing
AMMC	Aluminium Metal Matrix Composite
ANFIS	Adaptive Neuro-Fuzzy Inference System
ANN	Artificial Neural Network
ANOVA	Analysis of Variance
AR/VR	Augmented and Virtual Reality
ARC	Airworthiness Review Certificate
ARPANET	Advanced Research Project Agency Network
AWS	Amazon Web Services
BSE	Back Scattered Electrons
CAMP	Continuous Airworthiness Maintenance Program
CBM	Conditioned-Based Monitoring
CBS	Cyber-Physical Systems
CM	Corrective Maintenance
Ct	Charpy Strength in Joule
D	Derivative Term
DED	Directed Energy Deposition
DL	Deep Learning
DMLS	Direct Metal Laser Sintering
DOF	Degree of Freedom
DP	Dual Phase
DTR	Decision Tree Regression
%E	Percentage of Elongation
EASA	European Aviation Safety Authority
EBM	Electron Beam Melting
EBSD	Electron Backscatter Diffraction
EDM	Electro Discharge Machining
EDS	Energy-Dispersive X-Ray Spectroscopy
EDS	Energy Dispersive Spectroscopy
EDX	Energy Dispersive X-Ray Analysis
EPW	Electromagnetic Pulse Welding
ETD	Everhart–Thornley Detector
EW	Explosive Welding
FAA	Federal Aviation Authority
FDM	Fused Deposition Modelling
FFNN	Feed Forward Neural Network
FIB	Focused Ion Beam
FSAM	Friction Stir Additive Manufacturing
FSP	Friction Stir Processing
FSW	Friction Stir Welding
FW	Forge Welding

GCP	Google Cloud Platform
GMAW	Gas Metal Arc Welding
GPR	Gaussian Process Regression
GPUs	Graphics Processing Units
GRA	Grey Relational Analysis
GRC	Grey Relational Coefficient
GRG	Grey Relational Grade
HAZ	Heat Affected Zone
HB	Higher The Best
HT	Holding Time
I	Integral Term
I 4.0	Industry 4.0.
IIoT	Industrial Internet of Things
IMCs	Intermetallic Compounds
IoT	Internet of Things
IRI	Innovation Readiness Index
JIT	Just In Time
LB	Lower the Best
LC	Least Count
LGBR	Light Gradient Boost Regression
LMAIS	Liquid Metal Alloy Ion Sources
LMIS	Liquid Metal Ion Sources
LOM	Laminated Object Manufacturing
LR	Linear Regression
M2M	Machine-To-Machine Communication
Mg	Magnesium
MIG	Metal Inert Gas
ML	Machine Learning
MQL	Minimum Quantity Lubrication
MRO	Maintenance, Repair, and Overhaul
NB	Nominal the Best
NP	Number of Pass
NZ	Nugget Zone
OEE	Overall Equipment Effectiveness
OpSpecs	Operation Specifications
P	Proportional Term
PAC	Pierre Audo in Consultants
PBF	Powder Bed Fusion
PD	Proportional-Derivative
PdM	Predictive Maintenance
PI	Proportional-Integral
PID	Proportional-Integral-Derivative
PLC	Programmable Logic Controller
PLCs	Programmable Logic Controllers
PM	Preventative Maintenance
PR	Polynomial Regression

PSS	Product-Service System
QPSO	Quantum Particle Swarm Optimization
RFR	Random Forest Regression
RMR	Rate of Material Removal
ROS	Robotic Operating System
RSW	Resistance Spot Welding
RTV	Robot Transport Vehicle
RUL	Remaining Useful Life
S	Tool Rotation Speed
SCADA	Supervisory Control and Data Acquisition
SCARA	Selective Compliant Assembly Robot Arm
SDL	Selective Deposition Lamination
SEM	Scanning Electron Microscope
SLM	Selective Laser Melting
SLS	Selective Laser Sintering
SMEs	Small and Medium-Sized Businesses
SML	Supervised Machine Learning
SVM	Support Vector Machine
SVR	Support Vector Regression
T	Temperature
TEM	Transmission Electron Microscope/Microscopy
TIG	Tungsten Inert Gas
TMAZ	Thermo Mechanical Affected Zone
TS	Traverse Speed
Ts	Tensile Strength in Mpa
UAM	Ultrasonic Additive Manufacturing
USW	Ultrasonic Welding
UTS	Ultimate Tensile Strength
WOC	Width of Cut
XGBR	Xgboost Regression
XRD	X-Ray Diffraction
YS	Yield Strength

1 Enhancement of Manufacturing Sector Performance with the Application of Industrial Internet of Things (IIoT)

Pravatnallini Chhotaray, Bikash Chandra Behera and Bikash Ranjan Moharana
C. V. Raman Global University, Bhubaneswar, India

Kamalakanta Muduli
Papua New Guinea University of Technology, Lae, Papua New Guinea

Fono-Tamo Romeo Sephyrin
Tickle College of Engineering, University of Tennessee, Knoxville, Tennessee

CONTENTS

1.1 INTRODUCTION

In today's modern manufacturing environment, there is an increasing emphasis on using data and intelligence to improve efficiency, quality, and overall performance.

DOI: 10.1201/9781003346623-1

1

The use of advanced technologies such as the Internet of Things (IoT), artificial intelligence (AI), machine learning, and data analytics have revolutionized the way manufacturers approach their operations. Manufacturers can use data analytics to collect, analyze, and interpret large amounts of data generated by their production processes, supply chains, and customer interactions. This can help identify areas for improvement, optimize production workflows, and reduce waste, leading to significant cost savings. Additionally, the use of AI and ML can help predict and prevent equipment breakdowns, improve product quality, and reduce downtime. These technologies can also be used to develop predictive maintenance schedules, ensuring that maintenance is only carried out when necessary, rather than on a fixed schedule, reducing costs and downtime. Furthermore, the use of intelligent systems such as robotics and automation has greatly increased productivity and reduced the potential for errors in the manufacturing process. These systems can perform complex tasks with high precision, speed, and consistency, allowing manufacturers to produce products more efficiently and cost-effectively. Overall, intelligence and a focus on data have become crucial for modern manufacturers looking to stay competitive and improve their operations. The ability to collect and analyze data in real-time can help manufacturers make informed decisions, reduce costs, increase efficiency, and ultimately improve their bottom line (Kusiak, 2018; Das et al., 2023). In this day and age of the IIoT, a small production unit may be viewed as a powerful linked industrial system that contains materials, components, equipment, tools, and inventory and communicates with one another. This refers to connected devices, sensors, and other equipment that may be networked in an industrial context to offer remote access, effective monitoring, improved data collection, analysis, sharing, and other similar capabilities. The IoT has evolved into the next level of technology known as IIoT. The IIoT has the potential to bring benefits in all areas, including product delivery, work flow for employees, maintenance, and logistics suppliers. The IIoT makes it possible for manufacturers to digitalize practically every aspect of their operations (Boyes et al., 2018; Pizam, 2017; Verdouw et al., 2016). The analogue world is becoming digitized, which will eventually lead to the interconnection of everything. The proliferation of highly intelligent gadgets and technology has made it possible for humans to maintain continual contact at any time and in any location. The IIoT is linking the physical world of sensors, devices, and machines with the internet. Additionally, by using advanced analytics through software, the IIoT is converting huge amounts of data into powerful new insights and knowledge. The IIoT is the use and expansion of the IoT in industrial sectors and applications, with a significant emphasis on machine-to-machine communication (M2M), large data sets, and machine learning (Boyes et al., 2018; Malik et al., 2021). It is a network of connected devices and machines that collect and transmit data to provide insights into the performance of industrial processes. These intelligent machines can help businesses and industries to improve their operations by providing accurate, consistent, and real-time data that can be used to optimize processes, reduce downtime, and improve efficiency. By using sensors, actuators, and other devices to capture data, the IIoT can provide valuable insights into everything from supply chain management to manufacturing processes. These insights can help businesses to identify inefficiencies, streamline operations, and reduce costs. One of the key advantages of the IIoT is that it enables machines to communicate with each other and with humans in real time,

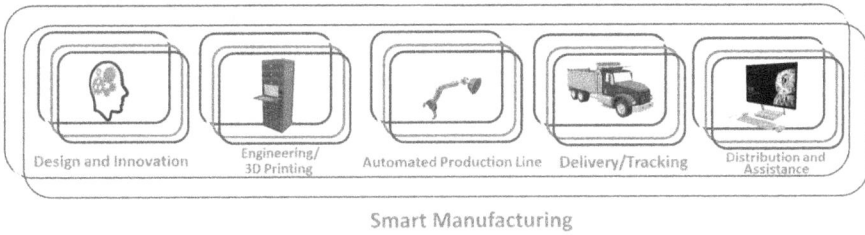

Smart Manufacturing

FIGURE 1.1   Industry 4.0 (manufacturing and digital innovation).

providing immediate feedback on performance and allowing for quick adjustments and interventions as needed. This can help to prevent equipment failures, reduce downtime, and improve overall reliability. Furthermore, these data can be used to assist businesses in completing tasks in less time, resolving problems, saving money, and supporting business intelligence efforts. The IIoT is beneficial to the operational environments of the industry in two different ways. Firstly, it integrates smart connected machines and manufacturing assets with the wide-spread enterprise, which helps to create a more efficient and flexible environment. Secondly, it improves asset performance through the positioning of cost-effective wireless sensors, simple cloud connectivity, and data analytics. The collection of data from a variety of fields and their subsequent transformation into information that can be acted upon in real time leads to improvements in business operations.

The aim of this chapter is to cover the basics of IIoT technology, its benefits, features, and its impact on smart manufacturing (Figure 1.1). Additionally, it intends to identify significant applications of IIoT in the manufacturing industry. By achieving these objectives, the chapter will provide readers with insightful information on the current and potential applications of IIoT in the manufacturing industry.

1.2   IIoT AND IT'S NEED IN MANUFACTURING

IoT stands for the "Internet of Things," which is a network of physical devices, vehicles, home appliances, and other items that are embedded with sensors, software, and network connectivity. These devices can exchange data with other devices and systems over the internet, enabling them to collect and transmit data and perform actions based on that data. IoT technology has a wide range of applications, from smart homes and cities to industrial automation and healthcare. It is an important technology for enabling greater connectivity, automation, and efficiency in various industries and sectors. The Internet of Things (IoT) can be classified into different categories based on their application and use case. The three main categories of IoT are:

Commercial IoT: This category includes IoT devices and systems that are used in commercial settings, such as retail stores, offices, and public spaces. Examples of commercial IoT devices include smart cameras, sensors, and digital signage.

Industrial IoT (IIoT): This category includes IoT devices and systems that are used in industrial settings, such as factories, manufacturing plants, and energy plants. Examples of IIoT devices include industrial sensors, programmable logic controllers (PLCs), and industrial robots.

Embedded IoT: This category includes IoT devices and systems that are integrated into other products or devices, such as smart appliances, wearable devices, and smart home systems. Examples of embedded IoT devices include smart thermostats, smart watches, and connected cars.

Each category of IoT has its own unique challenges and requirements, and they are often designed with specific use cases in mind. However, they all share the common goal of connecting devices and data in order to improve efficiency, productivity, and convenience (Dzaferagic, Marchetti, and Macaluso, 2022; Malik et al., 2021).

The phrase "**Industrial internet of things**" is frequently used in the industrial sectors as part of a digital transformation. This transition integrates key assets, powerful predictive and prescriptive analytics, and contemporary industrial workers. It is a network of devices used in industry that are connected to one another using communications technology to create systems that can monitor, collect, exchange, analyze, and give critical new insights. These systems are then used to assist industrial organizations in making better and faster business decisions. The IIoT allows for the optimization of manufacturing and industrial processes via the use of intelligent sensors and actuations. The various benefits of IIoT are depicted in Figure 1.2.

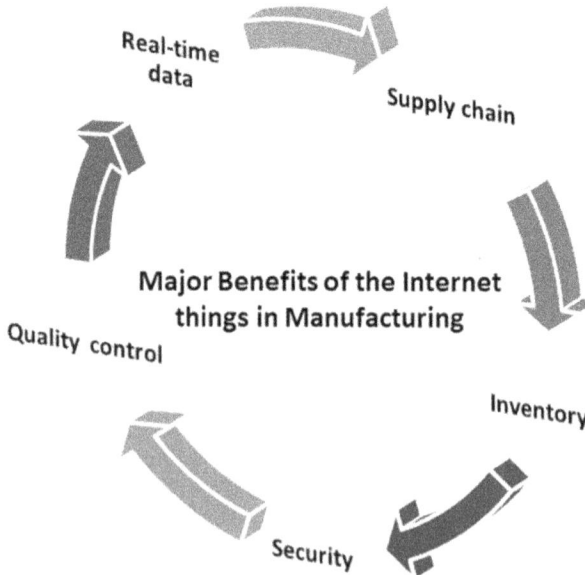

FIGURE 1.2   Major benefits of IIoT in manufacturing.

IIoT can bring numerous benefits to the industry, including:

- **Increased efficiency**: IIoT solutions enable the automation of various industrial processes, reducing manual intervention, and improving overall efficiency.
- **Predictive maintenance**: With IIoT solutions, industrial equipment can be monitored in real-time, providing insights into potential issues before they lead to downtime. This enables predictive maintenance, reducing unplanned downtime and associated costs.
- **Improved safety**: IIoT solutions can be used to monitor industrial environments for hazards, such as gas leaks or fires, allowing for quick response and improved safety.
- **Enhanced quality control**: IIoT solutions can provide real-time monitoring of the production process, enabling early detection of quality issues and reducing the likelihood of defective products.
- **Increased transparency**: IIoT solutions can provide stakeholders with real-time data on industrial processes, enabling better decision-making and increased transparency.
- **Improved supply chain management**: IIoT solutions can be used to track the movement of goods through the supply chain, providing real-time visibility into inventory levels and enabling better management of logistics.

Overall, linked IIoT solutions can bring numerous benefits to the industry, including increased efficiency, improved safety, and enhanced quality control. These benefits can translate into significant cost savings, increased productivity, and improved customer satisfaction.

1.3   THE DEFINING CHARACTERISTICS OF THE IIoT IN THE MANUFACTURING DOMAIN

The IIoT is the use of connected sensors, devices, and machines in manufacturing processes to collect data and improve productivity. Here are some of the defining characteristics of IIoT in the manufacturing domain:

- **Connectivity**: IIoT devices are connected to each other and to a central network or cloud-based system. This allows them to communicate with each other and share data in real-time.
- **Data Collection**: IIoT devices are equipped with sensors that collect data about various aspects of the manufacturing process, such as temperature, pressure, and speed. This data is then transmitted to a central system for analysis.
- **Data Analysis**: IIoT data is analyzed using advanced analytics tools, such as machine learning algorithms. This analysis can provide insights into the performance of machines, the efficiency of the manufacturing process, and potential areas for improvement.

- **Predictive Maintenance**: IIoT devices can be used for predictive maintenance, which involves using data analysis to predict when a machine will need maintenance or repair. This can help prevent downtime and reduce maintenance costs.
- **Automation**: IIoT devices can be used to automate various aspects of the manufacturing process, such as quality control, assembly, and packaging. This can help improve efficiency and reduce the need for human intervention.

Overall, the IIoT is transforming the manufacturing industry by providing real-time data and insights that can be used to improve productivity, reduce costs, and increase profitability. Enhancing the entire performance of the manufacturing sector may be accomplished with the help of the industrial internet of things (IIoT), which includes a variety of qualitative and adaptable characteristics. The characteristics and components of quality, such as the reduction of errors throughout the process, the utilization of intelligent and digital work equipment and stations, the cost reductions involved, the safety issues, etc. Also taken into consideration are important inputs from the IIoT toward the development of a successful and efficient manufacturing culture. The Industrial Internet of Things is a network that consists of important assets, potent predictive and prescriptive analyses, and modern industrial workers. It is a vast network of various types of industrial equipment connected to one another through various forms of communication technology that ultimately leads to systems. The most recent developments in the IoT industry may be advantageous both for entry-level workers and to offer new opportunities. By receiving vital information in real time, managers are able to determine whether or not the machinery in their charge requires maintenance and improve their ability to prepare for it. One of the many advantages for manufacturing is the method in which the IoT improves operational efficiency. Sensors that are linked are able to recognize the likely failure and trigger a request for an engineer to repair it (Cheng et al., 2018).

IoT is utilized by manufacturers to integrate dynamic, competent, and automated production activities into their products. These operations have autonomous maintenance schedules. Therefore, improved maintenance procedures are triggered, which result in substantial cost savings, a reduction in the number of machine failures, and an increase in the life of the machine. The Internet of Things in industry has the potential to dramatically transform manufacturing. IIoT may be able to aid in reducing the costs associated with the process through the automation and simplification of the process control plan. Businesses are able to save time and costs by automating the process of confirming the quality of crucial variables, thanks to sensors that can be connected to the internet of things to ease the procedure, rather than relying on human inspections (Jeschke et al., 2017).

The installation of IIoT sensors is being done in order to make the administration of the facility more effective. The use of sensors in conjunction with software designed for IIoT enables the monitoring of temperature, vibrations, and other factors that may contribute to less-than-ideal working conditions. In addition, intelligent lighting and intelligent sensors improve visibility and control over a company's resources, such as its electricity, water, and fuel supplies, among other things.

Environmental sensors have the ability to monitor quality-critical circumstances and continuously provide management with information. In a pharmaceutical operation, for instance, maintaining the proper temperature may be absolutely necessary to ensure product quality. By utilizing IoT linked sensors, managers are able to monitor a variety of factors and are immediately notified if any of those variables exceed the required parameters (Ordieres-Meré, Villalba-Díez, and Zheng, 2019).

The majority of companies, in particular, are concentrating their efforts on IIoT in order to decrease expenses associated with performance, security, and operation. IIoT has become the Industry 4.0 solution that is most widely used by modern manufacturing companies in order to satisfy a wide range of criteria. It is utilized in supply chains to perform real-time monitoring of inventory levels and to activate the automatic replenishing of supplies by suppliers. The events that occur across the supply chain may be traced with ease and offer an inexhaustive picture of the inventory. In conjunction with data analytics provided by the IIoT, corporate platforms and software developed by IIoT companies examine trends of product and material usage. It gives businesses the ability to get rid of inefficiencies in their processes and reduce waste. Companies have been relying on technical solutions for a significant amount of time.

In order to achieve lean goals such as higher performance and improved operations, IIoT solutions, and especially industrial IoT automation, make it possible for combination systems to deliver the highest levels of performance and availability across all assets, product lines, facilities, and the whole organization. It is essential to have safe storage for the data obtained from various pieces of equipment and sensors in order to do analysis and put them to use. The use of information may be made more adaptable and effective thanks to the cloud, which in turn makes it possible for industrial businesses to run it. Intelligent networks are used by manufacturing companies to bridge the gap between their networks and the corporate networks of their customers.

This allows for a reduction in downtimes by enabling remote access to systems, which in turn provide the organization with conditions that are exact, dependable, and robust from the plant floor. The IIoT makes it possible for plant managers to keep tabs on all of the components required for production at the product level. With this method, it is possible to replenish the stock without restarting the production procedure. Inventory planning will be improved as soon as the IoT becomes operational, allowing for more accurate long-term planning and improved forecasting of the amount of time required for certain resources to become obsolete (Lu, Witherell, and Jones, 2020). Through the use of the industrial internet, firms are able to move toward a predictive maintenance plan that is more proactive. It gives businesses the ability to monitor their machinery and equipment in real time and schedule maintenance at the exact moment the machine needs it.

This provides visibility in real time into all of the production processes that are involved in manufacturing organizations. In the past, it has been difficult to compare the levels of productivity and the quality of the items produced at each factory. Using these tools, companies can now collect data from several plants and analyze it all at once. By forecasting problems that may arise during the production cycle, it assists businesses in shifting away from a reactive management style and toward a more proactive one. The Internet of Things provides specific data to industrial companies on all devices used in production. The IIoT refers to autonomously communicating devices. Businesses have taken on an increasingly large number of goods in the man-ufacturing of IoT products to minimize bottlenecks while producing, to avoid exces-sive loads that might deteriorate equipment, and to improve productivity and product quality and consistency. These goals have led to an increase in the number of goods being manufactured. IIoT can disrupt corporate management in manufacturing. This offers new opportunities, new processes, and new things to purchase. The benefits of utilizing IIoT may be better understood by organizations with the aid of solutions that cover testing and training in their entirety. It is able to immediately analyze their successes and failures, as well as estimate the worth and degree of their disruptive potential.

The Industrial Internet of Things makes use of networked sensors and intelligent devices to collect data in order to advance AI and do predictive analysis immediately on the manufacturing floor. In the complete digitalization of the supply chain, also known as the digital supply chain, it is abundantly obvious that retail and customer concerns are directly influenced by IIoT as it plays a pivotal role in many subsystems and the value chain of the massive ecosystem. Other plants may improve their pro-ductivity, preventative maintenance, and data-driven decision making with the help of IoT-enabled operations and services.

1.4   MANUFACTURING IIoT FEATURES

In the manufacturing domain, IIoT enables the collection and analysis of data across the entire manufacturing process to improve efficiency, productivity, and safety. Here are some collective features (Figure 1.3) of IIoT for the manufacturing domain:

- **Real-time monitoring**: IIoT provides real-time monitoring of machines and processes, allowing manufacturers to identify problems as they occur and take corrective action immediately.
- **Predictive maintenance**: IIoT enables predictive maintenance, where machines and equipment are monitored continuously to detect potential problems before they become major issues. This helps to reduce downtime and improve overall equipment effectiveness (OEE).
- **Remote access and control**: IIoT allow remote access and control of machines and equipment, enabling manufacturers to monitor and adjust operations from anywhere in the world.
- **Improved inventory management**: IIoT helps to improve inventory management by providing real-time data on inventory levels, usage, and demand. This helps manufacturers to optimize their inventory levels, reduce waste, and improve supply chain efficiency.
- **Increased efficiency and productivity**: IIoT enable manufacturers to opti-mize their production processes, reduce waste, and improve overall effi-ciency and productivity.
- **Enhanced safety**: IIoT helps to enhance safety in the manufacturing pro-cess by providing real-time data on the condition of machines and equip-ment, as well as identifying potential safety hazards.

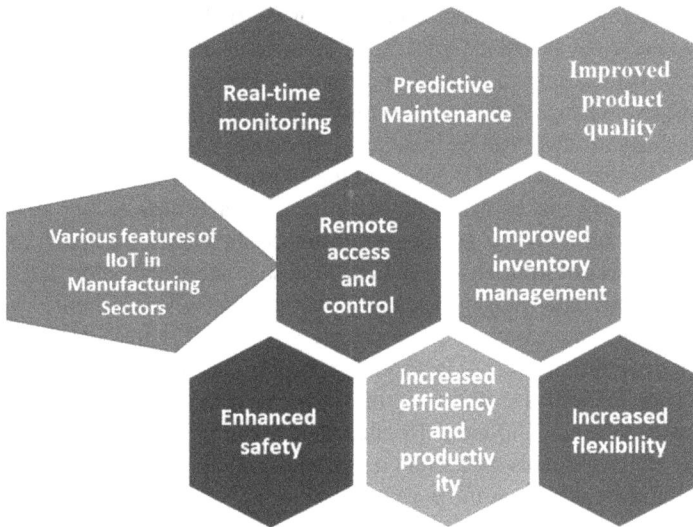

FIGURE 1.3   Various of features of IIoT in manufacturing sectors.

- **Data analytics**: IIoT generates a vast amount of data, which can be analyzed to gain insights into manufacturing processes, identify areas for improvement, and optimize performance.
- **Supply chain visibility**: IIoT enables supply chain visibility by providing real-time data on the movement of goods and materials through the supply chain, enabling manufacturers to optimize their supply chain operations.
- **Improved product quality**: IIoT helps to improve product quality by enabling manufacturers to monitor and control production processes more closely, ensuring that products meet quality standards.
- **Increased flexibility**: IIoT enables manufacturers to be more flexible and responsive to changing customer demands and market conditions by providing real-time data on production processes and inventory levels.

1.5   MAJOR APPLICATIONS OF IIoT IN MANUFACTURING INDUSTRIES

IIoT has become a crucial technology for today's manufacturing industries because of its ability to help businesses increase their productivity, lower their costs, and improve the quality of their products. The manufacturing industry is always being put under pressure to reduce overall production and operational expenses while simultaneously increasing the efficiency with which it produces environmentally friendly products. The importance of the IIoT is rapidly growing, while organizations are placing a significant amount of weight on the advantages offered by technology. It opens up a wide range of opportunities, particularly in the commercial and industrial spheres. The current level of assets and the quality of those assets determine the worth of a manufacturer. Connected devices for the IIoT make it possible for manufacturers to carry out automatic tracking of their assets and assist in the formation of relationships with partners and suppliers. The IIoT will make it possible for manufacturers to monitor changes in demand in real time. This enables businesses to respond more effectively to sudden shifts in demand. Because of this, the management of the warehouse, the distribution of assets, and the system of the supply chain will all be improved. The IIoT makes data analysis a practical and time-saving possibility. A continuous evaluation of the leading performance indicators for health and safety, such as the number of injuries, the number of absences for both short and long periods of time, the disease rate, and the number of near misses, is therefore possible in order to guarantee improvements in working conditions. It is possible to take prompt action to reduce the incidence of accidents such as labelling indications. The IIoT is able to keep track of the utilization of resources such as power, fuels, water, and so on. Sensors connected to the IoT collect data about products at various phases of the product life cycle. This data refers to the whole composition of the raw materials that were utilized, as well as the working conditions, waste, transit effect, and other factors that had an effect on the completed items. In addition to this, when the IoT device is integrated into the final product, information on the customer's perception of how they are using the product may be supplied. Businesses are able to collect crucial information about how their products are utilized and the number of people that buy them when they include IIoT into their products and packaging. The IIoT can be applied in a variety of ways in the manufacturing sector, including the following:

- **Monitoring the Health of Manufacturing Equipment**: IIoT sensors can be used to monitor the health of manufacturing equipment and forecast when maintenance is required. This helps to reduce downtime and increases efficiency.
- **Control of Quality**: Sensors connected to an IIoT system can keep an eye on the production process in real time, allowing for early detection of quality concerns before they balloon into more serious ones. This helps to reduce waste while also improving the quality of the product as a whole.
- **Optimization of the Supply Chain**: The IIoT can assist in the optimization of the supply chain by giving real-time data on inventory levels, production schedules, and delivery timeframes. This data can be used to make improvements. This has the potential to assist manufacturers in lowering their costs and enhancing their delivery times.
- **Energy Management**: IIoT sensors can monitor energy usage in manufacturing plants and discover potential for energy savings. This is an important part of energy management. This can assist manufacturers in lowering their overall energy expenses and reducing their overall carbon impact.
- **Asset Tracking**: IIoT sensors can be used to track the position and status of manufacturing assets such as raw materials, finished goods, and machinery. The operations of manufacturers can be optimized with the help of this, which also helps reduce the danger of loss or theft.

The literature about IIoT in Manufacturing is set forth in Table 1.1.

TABLE 1.1

Significant Application of IIoT in the Field of Manufacturing

Sl. No	Area of Applications	Summary	References
1	Improvement productivity	The primary reasons for implementing IIoT technologies are for the purposes of optimizing, enhancing productivity along the various internal value chains, while simultaneously improving operational efficiency. The IIoT is gaining popularity in many industrial settings and is increasingly being implemented in a variety of supply chain applications across industries, including manufacturing, automotive, pharmaceutical, electrical, high technology, and food-beverage supply chains. The benefits of this technology are readily available to a large number of workers. The IIoT makes use of a variety of different solutions, applications already in use, and technological advancements.	Mahbub (2020); Lin, Lan, and Huang (2019); Rosales, Deshpande, and Anand (2021)
2	Intelligent factories	IIoT enabled smarter factories and system integration for manufacturers. This helps advanced analysts spot patterns and make informed decisions. Systems integration helps industries manage energy, inventory, and supply chains. Industrial automation, central monitoring, predictive asset maintenance, cost reduction, resource optimization, profitability, and operating efficiency result. IoT-enabled smart homes, buildings, cars, and more could change our lives.	Javaid et al. (2021); Kavitha et al. (2021); Suthar and Peter He (2021)
3	Reduction in manufacturing expenses	Integrated systems can assist manufacturers in boosting production while reducing errors. By integrating various systems, manufacturers can streamline their processes, eliminate redundancies, and increase efficiency. Moreover, one of the key benefits of the Industrial Internet of Things (IIoT) is the ability to reduce costs associated with production inaccuracies and flaws. Digital twins are one of the features that can be used by manufacturers to solve a variety of issues using the IIoT. Digital twins create a virtual replica of the product that is currently under development, allowing manufacturers to collect data from each individual unit using sensors, including information about the entire operating mechanism of their equipment and the desired output. With the data obtained from the digital replica, management can evaluate the efficacy of the system as well as its level of precision, helping to identify potential bottlenecks in the product and build a more effective version of the product. This can help manufacturers reduce costs associated with production errors and increase the efficiency of their production processes, ultimately leading to greater profitability.	Qiu et al. (2020); Liu et al. (2020); Kharchenko et al. (2020)

(Continued)

TABLE 1.1 (Continued)

Sl. No	Area of Applications	Summary	References
4	Implementing Smart supply chain	IIoT devices can track goods and assets in real-time, allowing for accurate and timely delivery, reducing the likelihood of loss or theft, and improving inventory management. IIoT sensors can monitor the condition of equipment and machinery, allowing for predictive maintenance and reducing the likelihood of costly downtime. IIoT technology can provide real-time updates and data on all aspects of the supply chain, from inventory levels to production schedules, enabling greater visibility and transparency across the entire supply chain. IIoT-enabled machines and equipment can operate autonomously, reducing the need for human intervention and increasing efficiency. IIoT devices can collect and analyze data on various aspects of the supply chain, allowing for optimization of processes, identifying bottlenecks, and making more informed decisions. Overall, IIoT technology can greatly improve the efficiency and effectiveness of supply chain management, leading to reduced costs, increased productivity, and improved customer satisfaction.	Lin and Hwang (2019); Hu (2015); Arnold, Kiel, and Voigt (2017)
5	Real time inspection and information	Real-time inspection of IIoT can help businesses improve their operational efficiency, reduce costs, and enhance safety and quality in their industrial processes. One of the benefits of IIoT solutions is that they can provide real-time information on the status of various processes, such as the production of goods or the delivery of services. This information can be used to make informed decisions and respond quickly to changing circumstances. In addition, some IoT suppliers integrate their solutions with firms' current ERP (Enterprise Resource Planning) systems. This integration eliminates the need for manual operating paperwork, reducing the risk of errors, and streamlining processes. By automating processes and integrating systems, businesses can improve their overall efficiency and productivity. Finally, IoT solutions can also help businesses track the delivery of raw materials and equipment. This allows businesses to manage their inventory more effectively and ensure that they have the necessary resources to meet demand. Overall, IoT solutions offer many benefits to businesses, including increased efficiency, cost savings, and improved customer experience.	Fraile et al. (2018); Mourtzis, Vlachou, and Milas (2016); Park et al. (2020)

6	Simplification of the flow of manufacturing	IoT devices can have several uses in production facilities monitoring development cycles and managing inventory automatically. IoT devices can simplify the production flow and improve efficiency. IoT devices can also track worldwide inventory, monitor the supply chain, and provide meaningful estimations of available resources, helping industries optimize their production processes. Moreover, IoT devices can eliminate the need for manual documentation and implement Enterprise Resource Programs, improving accuracy and reducing errors. With IoT devices, management departments across different channels can have visibility into production processes and stakeholders can review progress in real-time. Overall, IoT can have a significant impact on the production flow and help industries achieve greater efficiency and productivity.	Mekid and Akbar (2019); Batalla (2020); Park and Jeong (2020)
7	Safety of worker	IoT has the potential to significantly improve worker safety in industrial settings. By connecting equipment and personnel through IoT devices, valuable insights can be gained into asset management and employee safety. For example, sensors can be placed on machines to monitor temperature, pressure, and other factors that could pose a safety risk to workers. IoT devices can also track the location of personnel within a facility, enabling employers to quickly locate employees in case of an emergency.	Sahu, Sahu, and Sahu (2020); Mayer, Tantscher, and Bischof (2020); Khan et al. (2020)
8	Security	IIoT devices can also be used to improve physical security in industrial settings. By monitoring machines and equipment in real-time, IIoT devices can detect signs of wear and tear and predict potential malfunctions, allowing for preventive maintenance to be carried out before an accident occurs. Furthermore, IIoT devices can help ensure that employees comply with safety requirements. For example, sensors can be used to monitor worker behavior and provide feedback on compliance with safety protocols. This can help identify areas where additional training or safety measures may be needed, reducing the risk of accidents and injuries. Another advantage of IIoT monitoring is that it can be implemented in older facilities without interfering with existing industrial control networks. This means that even factories with older equipment and infrastructure can benefit from the safety and productivity improvements that IIoT can provide.	Sengupta, Ruj, and Das Bit (2020); Lin et al. (2018); Wiesner and Wuest (2020); Candell et al. (2020)

(Continued)

TABLE 1.1   (Continued)

Sl. No	Area of Applications	Summary	References
9	Inspection of product quality	IIoT devices can provide manufacturers with comprehensive product data at various phases of the product cycle using thermal and video sensors. This enables manufacturers to monitor the quality of the products at every step of production, ensuring that their attributes are specified. By using IoT devices to monitor equipment and the results of each manufacturing stage, manufacturers can be more proactive in identifying and addressing quality issues. They can use customized end-user dashboards provided by IoT service providers to examine the outcomes of smart monitoring thoroughly. This can help manufacturers evaluate the cost, efficiency, and carbon impact of other resources so that they can explore alternative production processes. In this way, IIoT devices can help manufacturers improve the quality of their products and reduce waste and inefficiency. By providing real-time data on production processes and enabling manufacturers to quickly identify and address quality issues, IIoT can help manufacturers achieve greater efficiency and productivity while also reducing their environmental impact.	Biurrun et al. (2020); Iyenghar, Sundharam, and Pulvermueller (2020); Cha et al. (2019)
10	Decision making using IIoT	IoT can significantly enhance the decision-making process in an organization by providing real-time data on the performance and insights of various devices within the network. This data can help organizations identify inefficiencies, potential issues, and opportunities for improvement. In light industrial IIoT applications, such as meters, IoT can provide insights on energy consumption and usage patterns, which can help organizations optimize their energy usage and reduce costs. In heavy industry applications, IoT devices can monitor equipment performance and detect issues before they become critical, enabling organizations to take proactive measures to prevent downtime and reduce maintenance costs. IoT can also improve efficiency, accuracy, and safety processes through automation. For example, IoT-enabled machines can automatically adjust their operations based on real-time data, reducing the need for manual intervention and improving overall efficiency. IoT sensors can also detect changes in temperature, flow, or volume, and trigger automatic responses to maintain optimal conditions.	Gupta (2021)

1.6 LIMITATIONS

The deployment of IIoT is accompanied by its own unique set of restrictions and difficulties. Figure 1.4 shows a list of its main limitations, some of which include the following:

Cybersecurity: One of the primary concerns with IIoT is the security of connected devices and the data they generate. As more devices are connected, the attack surface increases, making it more challenging to protect against cyber threats.

Integration challenges: Integrating IIoT devices into existing systems can be a complex process. It requires a deep understanding of the existing infrastructure and the ability to identify and overcome potential compatibility issues.

Cost: Implementing IIoT can be costly, particularly for small-scale industries that may not have the budget to invest in the necessary technology and infrastructure.

Lack of expertise: There is a shortage of skilled professionals who can design, deploy, and maintain IIoT systems. This can make it challenging for organizations to adopt the technology.

To address these challenges, organizations need to carefully consider their goals and objectives before implementing IIoT. They need to evaluate the costs, benefits, and potential risks of the technology and develop a comprehensive strategy that takes into account all of these factors. Additionally, organizations need to work with trusted technology partners and implement robust security measures to ensure the integrity of their systems and data.

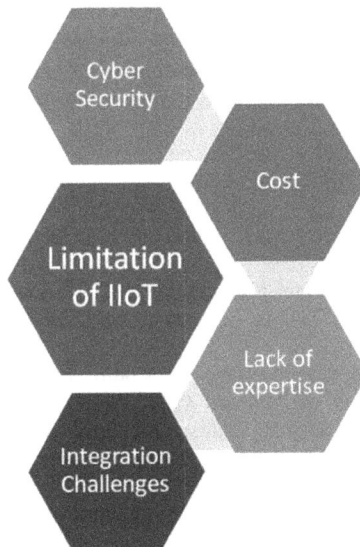

FIGURE 1.4   Main limitations of IIoT.

1.7 FUTURE SCOPE

In the future, IIoT will revolutionize the manufacturing industry by providing real-time data and insights that can be used to optimize production processes and improve overall efficiency. By collecting and analyzing data from sensors and other connected devices throughout the manufacturing process, manufacturers can identify bottle-necks, reduce downtime, and improve quality control. In addition, as automation technologies continue to advance, employees in the manufacturing industry will likely see their roles evolve to focus more on data analysis, problem-solving, and other tasks that require a human touch. This shift will require workers to be more tech-savvy and adaptable than ever before, but it also has the potential to create new opportunities for career growth and development.

1.8 CONCLUSIONS

Modernizations in the manufacturing sector's real-time efficiency are mostly the result of shifts in production methods, supply chains, robotic facilities, embedded technologies, and networked machinery. These measures not only minimize risk but also encourage fresh ideas and innovation. The use of reliable, low-cost, and net-worked sensors in automated factories leads to better output. Soon, the IIoT will be useful for things like self-diagnosis, mending broken industrial machinery, and ensuring the safety of production lines. Efficiency gains may be realized in a number of areas, including maintenance time, asset utilization, cost savings, worker output, the capacity to quantify outcomes, the quality of the final product, the ability to push for further improvements in efficiency, and finally, energy consumption. By reducing overhead costs and creating new revenue streams, businesses benefit. Technology like this might help reduce product development times by making production and logistics processes more streamlined. The Industrial Internet of Things (IIoT) makes it possible to learn about and discuss cutting-edge developments in manufacturing. By the use of sensors, data processing, and cameras, machines may now be equipped to anticipate when they may break down thanks to the IIoT. Industrial organizations will soon be able to use this technology to make faster and more informed decisions regarding their operations.

REFERENCES

Arnold, Christian, Daniel Kiel, and Kai-Ingo Voigt. 2017. Innovative business models for the industrial Internet of Things. *BHM Berg- und Hüttenmännische Monatshefte* 162 (9):371–381.

Batalla, J. M. 2020. On analyzing video transmission over wireless WiFi and 5G C-band in harsh IIoT environments. *IEEE Access* 8:118534–118541.

Biurrun, Aitor, Imanol Picallo, Hicham Klaina, et al. 2020. Implementation of a WSN-based IIoT monitoring system within the workshop of a solar protection curtains company. *Engineering Proceedings* (1).

Boyes, Hugh, Bil Hallaq, Joe Cunningham, and Tim Watson. 2018. The industrial internet of things (IIoT): An analysis framework. *Computers in Industry* 101:1–12.

Candell, R., M. Kashef, Y. Liu, K. Montgomery, and S. Foufou. 2020. A graph database approach to wireless IIoT Workcell performance evaluation. Paper read at *2020 IEEE International Conference on Industrial Technology (ICIT)*, 26–28 Feb. 2020.

Cha, Seungwoo, Jae-Hyun Yun, Youho Myong, and Hyung-Ik Shin. 2019. Spasticity and preservation of skeletal muscle mass in people with spinal cord injury. *Spinal Cord* 57 (4):317–323.

Cheng, Jiangfeng, Weihai Chen, Fei Tao, and Chun-Liang Lin. 2018. Industrial IoT in 5G environment towards smart manufacturing. *Journal of Industrial Information Integration* 10:10–19.

Das, Rashmi Prava, Kamalakanta Muduli, Sonia Singh, Bikash Chandra Behera, and Adimuthu Ramasamy. 2023. Unveiling the Role of Evolutionary Technologies for Building Circular Economy-Based Sustainable Manufacturing Supply Chain. In *Digital Transformation and Industry 4.0 for Sustainable Supply Chain Performance*, edited by S. S. Kamble, R. S. Mor, and A. Belhadi. Cham: Springer International Publishing.

Dzaferagic, M., N. Marchetti, and I. Macaluso. 2022. Fault Detection and Classification in Industrial IoT in Case of Missing Sensor Data. *IEEE Internet of Things Journal* 9 (11):8892–8900.

Fraile, F., T. Tagawa, R. Poler, and A. Ortiz. 2018. Trustworthy Industrial IoT Gateways for Interoperability Platforms and Ecosystems. *IEEE Internet of Things Journal* 5 (6):4506–4514.

Gupta, Vijay Prakash. 2021. Smart Sensors and Industrial IoT (IIoT): A Driver of the Growth of Industry 4.0. In *Smart Sensors for Industrial Internet of Things: Challenges, Solutions and Applications*, edited by D. Gupta, V. Hugo C. de Albuquerque, A. Khanna, and P. L. Mehta. Cham: Springer International Publishing.

Hu, Peng. 2015. A system architecture for software-defined industrial Internet of Things. Paper read at *2015 IEEE International Conference on Ubiquitous Wireless Broadband (ICUWB)*.

Iyenghar, P., S. M. Sundharam, and E. Pulvermueller. 2020. Integrated Performance Tuning of an IIoT Digital Twin: Work-in-Progress. Paper read at *2020 International Conference on Embedded Software (EMSOFT)*, 20–25 Sept. 2020.

Javaid, Mohd, Abid Haleem, Ravi Pratap Singh, and Rajiv Suman. 2021. Significance of Quality 4.0 towards comprehensive enhancement in manufacturing sector. *Sensors International* 2:100109.

Jeschke, Sabina, Christian Brecher, Tobias Meisen, Denis Özdemir, and Tim Eschert. 2017. Industrial Internet of Things and Cyber Manufacturing Systems. In *Industrial Internet of Things: Cybermanufacturing Systems*, edited by S. Jeschke, C. Brecher, H. Song, and D. B. Rawat. Cham: Springer International Publishing.

Kavitha, R. J., T. Avudaiyappan, T. Jayasankar, and J. Arputha Vijaya Selvi. 2021. Industrial Internet of Things (IIoT) with Cloud Teleophthalmology-Based Age-Related Macular Degeneration (AMD) Disease Prediction Model. In *Smart Sensors for Industrial Internet of Things: Challenges, Solutions and Applications*, edited by D. Gupta, V. Hugo C. de Albuquerque, A. Khanna, and P. L. Mehta. Cham: Springer International Publishing.

Khan, W. Z., M. H. Rehman, H. M. Zangoti, M. K. Afzal, N. Armi, and K. Salah. 2020. Industrial internet of things: Recent advances, enabling technologies and open challenges. *Computers & Electrical Engineering* 81:106522.

Kharchenko, V., O. Illiashenko, O. Morozova, and S. Sokolov. 2020. Combination of Digital Twin and Artificial Intelligence in Manufacturing Using Industrial IoT. Paper read at *2020 IEEE 11th International Conference on Dependable Systems, Services and Technologies (DESSERT)*, 14–18 May 2020.

Kusiak, Andrew. 2018. Smart manufacturing. *International Journal of Production Research* 56 (1–2):508–517.

Lin, Chao, Debiao He, Xinyi Huang, Kim-Kwang Raymond Choo, and Athanasios V. Vasilakos. 2018. BSeIn: A blockchain-based secure mutual authentication with fine-grained access control system for industry 4.0. *Journal of Network and Computer Applications* 116:42–52.

Lin, Hsien-I, and Yu-Che Hwang. 2019. Integration of robot and IIoT over the OPC unified architecture. Paper read at 2019 International Automatic Control Conference (CACS).

Lin, Yu-Ju, Ci-Bin Lan, and Chin-Yin Huang. 2019. A Realization of Cyber-Physical Manufacturing Control System Through Industrial Internet of Things. *Procedia Manufacturing* 39:287–293.

Liu, Y., T. Dillon, W. Yu, W. Rahayu, and F. Mostafa. 2020. Missing Value Imputation for Industrial IoT Sensor Data With Large Gaps. *IEEE Internet of Things Journal* 7 (8):6855–6867.

Lu, Yan, Paul Witherell, and Albert Jones. 2020. Standard connections for IIoT empowered smart manufacturing. *Manufacturing Letters* 26:17–20.

Mahbub, Mobasshir. 2020. Comparative link-level analysis and performance estimation of channel models for IIoT (Industrial-IoT) wireless communications. *Internet of Things* 12:100315.

Malik, Praveen Kumar, Rohit Sharma, Rajesh Singh, et al. 2021. Industrial Internet of Things and its Applications in Industry 4.0: State of The Art. *Computer Communications* 166:125–139.

Mayer, Barbara, Dominik Tantscher, and Christian Bischof. 2020. From Digital Shop floor to Real-Time Reporting: an IIoT Based Educational Use Case. *Procedia Manufacturing* 45:473–478.

Mekid, Samir, and Usman Akbar. 2019. Configuration and Business Protocol of International Load Sharing of Manufacturing and its Challenges Under I4. 0 and IIoT. Paper read at ASME International Mechanical Engineering Congress and Exposition.

Mourtzis, D., E. Vlachou, and N. Milas. 2016. Industrial Big Data as a Result of IoT Adoption in Manufacturing. *Procedia CIRP* 55:290–295.

Ordieres-Meré, Joaquín, Javier Villalba-Díez, and Xiaochen Zheng. 2019. Challenges and Opportunities for Publishing IIoT Data in Manufacturing as a Service Business. *Procedia Manufacturing* 39:185–193.

Park, Byungjun, and Jongpil Jeong. 2020. A CPS-Based IIoT Architecture Using Level Diagnostics Model for Smart Factory. Paper read at *Computational Science and Its Applications – ICCSA 2020*, 2020//, at Cham.

Park, Kyu Tae, Yong Tae Kang, Suk Gon Yang, et al. 2020. Cyber Physical Energy System for Saving Energy of the Dyeing Process with Industrial Internet of Things and Manufacturing Big Data. *International Journal of Precision Engineering and Manufacturing-Green Technology* 7 (1):219–238.

Pizam, Abraham. 2017. The internet of things (IoT): The next challenge to the hospitality industry. *International Journal of Hospitality Management* 62:132–133.

Qiu, T., J. Chi, X. Zhou, Z. Ning, M. Atiquzzaman, and D. O. Wu. 2020. Edge Computing in Industrial Internet of Things: Architecture, Advances and Challenges. *IEEE Communications Surveys & Tutorials* 22 (4):2462–2488.

Rosales, Jonathan, Sourabh Deshpande, and Sam Anand. 2021. IIoT based Augmented Reality for Factory Data Collection and Visualization. *Procedia Manufacturing* 53:618–627.

Sahu, Anoop Kumar, Atul Kumar Sahu, and Nitin Kumar Sahu. 2020. A Review on the Research Growth of Industry 4.0: IIoT Business Architectures Benchmarking. *International Journal of Business Analytics (IJBAN)* 7 (1):77–97.

Sengupta, Jayasree, Sushmita Ruj, and Sipra Das Bit. 2020. A Comprehensive Survey on Attacks, Security Issues and Blockchain Solutions for IoT and IIoT. *Journal of Network and Computer Applications* 149:102481.

Suthar, Kerul, and Q. Peter He. 2021. Multiclass moisture classification in woodchips using IIoT Wi-Fi and machine learning techniques. *Computers & Chemical Engineering* 154:107445.

Verdouw, C. N., J. Wolfert, A. J. M. Beulens, and A. Rialland. 2016. Virtualization of food supply chains with the internet of things. *Journal of Food Engineering* 176:128–136.

Wiesner, Stefan, and Thorsten Wuest. 2020. *International Journal of Automation Technology* 11(1):4–16. DOI: 10.20965/ijat.2017.p0004ER.

2 Reliability Prediction Using Machine Learning Approach

Granville James Embia
PNG University of Technology, Lae, Papua New Guinea

Bikash Ranjan Moharana and Bikash Chandra Behera
C. V. Raman Global University, Bhubaneswar, Odisha, India

Noor Hafizah Mohmaed
Department of Mechanical Engineering, Universiti Malaysia, Perlis, Malaysia

Dillip Kumar Biswal
Krupajal Engineering College, Bhubaneswar, Odisha, India

Kamalakanta Muduli
PNG University of Technology, Lae, Papua New Guinea

CONTENTS

2.1   INTRODUCTION

The use of a diverse array of sensors and other devices that are connected to the internet has made it possible to generate extremely large data collections. Also, because there is so much data, there is a chance that some of it is wrong when it is used in applications or stages of applications. In its most basic form, machine learning (ML) involves making use of data that has been previously stored, mined, extracted, and inferred in order to provide results that have been "trained." The benefits of quality assurance brought about by ML are growing in significance as the number of applications that rely on it increases. Increasing overall consumer happiness across a broad range of product categories, accelerating sales and marketing cycles, increasing productivity and profitability, identifying high risks, and providing early warnings in risk management, forecasting events, and so on are some examples where ML has been successfully employed to predict the outcomes. ML helps to alleviate some of the challenges that are caused by rigid and inconvenient design by taking over tasks that can be hazardous for humans as well. Learning machines have the potential to achieve improved levels of dependability, quality, productivity, early warning, accessibility, and ease of maintenance. Also, it may be able to do output analysis in a shorter amount of time and convey results to stakeholders with greater ease.

ML is a catch-all term for different ways to use artificial intelligence. It is most often used in the information technology sector. There are many different kinds of ML algorithms, but the four primary categories are supervised, semi-supervised, unsupervised, and reinforcement (Sarker, 2021). The following is a list of the primary components that make up an algorithm for ML: datasets for training and testing; an objective function or loss function to optimize, such as a sum of squared errors or a likelihood function; an optimization method and a model for the data; and a model for the data (for example, linear, nonlinear, or nonparametric) (Xu and Saleh, 2021). It is an effective method for training AI models, which may help in the automation or optimization of complex systems such as robotics, activities involving autonomous driving, manufacturing, and supply chain logistics.

Based on the above facts, it was realized that there is a need to study the effects of ML and how quality control practices can be improved with the help of ML. More specifically, there is a need to study how ML helps with quality assurance and quality control in industry, such as with equipment maintenance, manufacturing processes, figuring out how long a part has left to work, etc. Deep learning (DL), a subfield of machine learning that was derived from artificial neural networks (ANN) and is distinguished by its several nonlinear processing layers (Li et al., 2019), is able to estimate an item's remaining useful life. The fact that the operating conditions may be very different in many different parts of the world makes it hard to predict RUL. Changes in the operating conditions can result in RUL calculations that are contradictory to the information provided by the manufacturer in the form of technical manuals and standards. Thus, the solution to problems of this nature may lie in the application of a strategy that is flexible and makes use of machine learning. The focus of this research is on the exploration of potential applications of ML to determine quality assurance and reliability in order to guarantee standards (specification limits) and customer satisfactions in (mass) production, manufacturing, marketing, transport, weather forecasts, and other areas of business. The study is conducted with the following objectives.

- To find out how ML techniques might help businesses in PNG and the Asia-Pacific predict how reliable their assets will be.
- To explore how useful, important, or cost-effective ML techniques might be for businesses in PNG and the Asia-Pacific.

2.2 PROBLEM STATEMENT AND OBJECTIVES

At the moment, most maintenance work in PNG is either CM or PM, which is expensive for the companies. Companies also report huge economic and man-hour losses owing to unreliable maintenance support systems. Further, most manufacturing organizations have to import their machine components from other countries. Hence, quite often, production is stopped once a machine component fails to work for a relatively longer period in comparison to other developed countries. Hence, a need for comprehensive analysis of the relevant literature to explore the opportunities to improve the performance of maintenance activities and enhance reliability prediction using ML was felt. The evaluation was conducted with the intention of assisting businesses located in PNG with the upkeep of their assets through the use of ML algorithms that anticipate the dependability of such assets. The standard process for reliability prediction is shown in Figure 2.1.

2.3 LITERATURE REVIEW

In this section of the article, we are going to look at some of the results that were drawn from prior research publications on this subject, as well as what ML algorithms those articles used. In the context of ML, "reliability" means the practice of using data analytics to predict the rate of deterioration or failure of an asset so that it can be fixed or replaced before it breaks down completely. ML approaches to predict reliability can be used in factories, car factories, and pretty much any other place that has assets with rotating or moving parts that wear out quickly and cost a lot to fix or

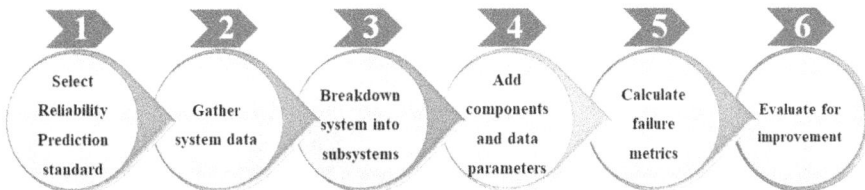

FIGURE 2.1 The reliability prediction process (Lees, 2012).

replace. These methods are applicable not just in automotive production facilities but also in manufacturing facilities in general. In recent years, machine learning algorithms have been used all over the world to forecast maintenance (Duchesne et al., 2020; Di Napoli et al., 2020; Wu et al., 2012). This trend began in the United States where engineers have proved that this approach has a greater accuracy rate when compared to the typical predictive maintenance strategy. The reason for this success is that this method can forecast failures in a machine much more accurately.

When it comes to the reliability of machines, PdM can present a number of challenges at times. This is especially true if there isn't enough data from machine operations that were done by trial and error over a certain amount of time. So, given that we live in an age where IT infrastructure is very important, it is possible to use engineering knowledge that is in line with the theory of MML to help predict how reliable equipment will be. Only a small number of businesses in PNG use this method to handle their maintenance tasks. This is in contrast to how widely the ML method for predicting reliability is used in the world's different business sectors. The analysis that follows is based on online articles written by authors who, like me, have done research on subjects related to the one I will be discussing. It is easy to see why predictive maintenance is the most effective method for preventing future breakdowns in the components of machinery used in a business. As a result, more attention needs to be drawn to it in order to guarantee that it will please all of the audiences who are involved. In recent years, data-based methods such as machine learning have been utilized to do predictive maintenance tasks, which ensure the equipment' continued reliability (Theissler et al., 2021; Eddarhri et al., 2022; Afshari et al., 2022). The subfield of AI known as ML enables computers to perform a variety of activities without being explicitly trained to do so. Although machine learning strategies can be broken down into four distinct categories—namely, unsupervised learning, semi-supervised learning, supervised learning, and reinforcement learning—in this article, I will focus on the broader topic of machine learning. This article from *Science Direct* explores the use of machine learning for predictive maintenance, focusing primarily on its application in the automotive sector. The concept that has been developed as a result of reading this essay is, in my opinion, applicable to all fields that make use of machines or components that move or rotate. According to the author, machine learning can be helpful in forecasting maintenance and, as a result, can improve the reliability of modern vehicles. This is due to the fact that all modern vehicles have a massive database. Machine learning has become a common practice in many manufacturing facilities, mobility solutions, and other areas because it helps predict maintenance, quality, safety, warranty, or plant facility monitoring and achieves the most important objective of every running organization, which is to save money and increase revenue. This is because machine learning helps achieve the most important goal of every organization. Despite the wealth of information contained in this document, the author made it abundantly clear that there is still a gap in knowledge that needs to be filled by additional research in the specific field of machine learning and predictive maintenance of automobiles. Conditioned-based monitoring (CBM) can be thought of as the basis for a decision that is made in predictive maintenance (PdM), also known as condition-based maintenance. It could be for certain components or parts, an assembly, a machine, or even the complete production line. In this case, a team of engineers will regularly keep an eye on it and figure out how long it will last and when it will break or need to be fixed or replaced. This strategy can help any company save a significant amount of money, making it a very cost-effective option. In particular, this study (Paolanti et al., 2018) presents the architecture of ML for PdM that is built on the Random Forest Method. Data was acquired using this way after a system was assessed. The results of the analysis were then compared to the results of the simulation tool using machine learning. Sensors, programmable logic controllers (PLCs), and other communication protocols were used to collect the data that was used for this study. The data was then analyzed with Azure Cloud Architecture, and the results successfully forecast a high degree of accuracy for various machine states. Reading this post gave me a much better understanding of how the ML approach may function by forecasting the reliability of any component by making use of technical expertise, and it was incredibly helpful to me. In situations in which it has been demonstrated to be more effective or to have more favorable effects than preventive and corrective maintenance activities, predictive maintenance is sometimes also referred to as the pro-active maintenance activity. However, it has some restrictions when it comes to the optimization of maintenance and the improvement of reliability (Yali Ren, 2021). Çınar et al. (2020) provide a summary of the maintenance classifications to act as an introduction to the topic. Because PdM has the potential to produce such features, it has proven to be one of the most promising strategies among other maintenance techniques that are capable of achieving such features. Other maintenance techniques that are capable of attaining such features include: Figure 2.2 depicts the maintenance characteristics that have been observed. Over the course of several years, ML has demonstrated that it is capable of overcoming the limitations of the conventional predictive maintenance

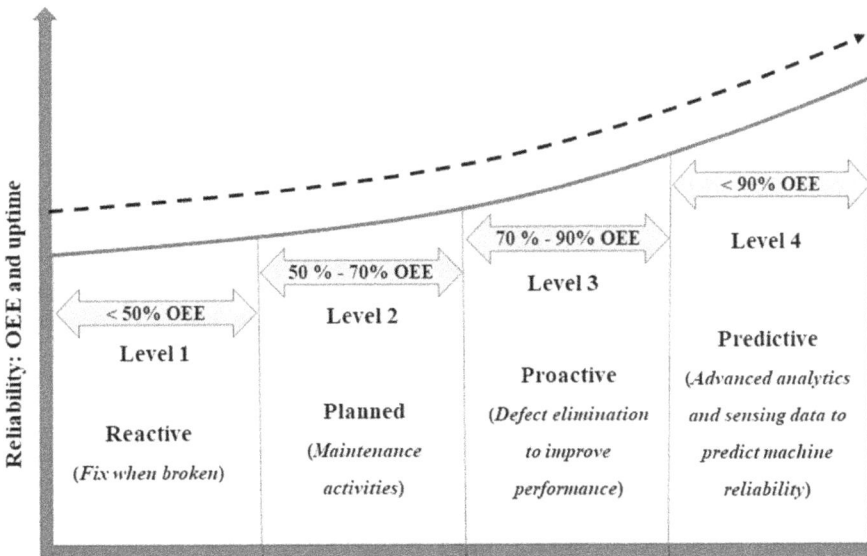

FIGURE 2.2   Different maintenance types (adapted from Çınar et al., 2020).

approach that was previously used. The field of predictive maintenance and reliability optimization is one in which machine learning has already demonstrated how useful it can be. The author discusses the benefits of machine learning in comparison with the traditional method, focusing specifically on the supervised and reinforcement learning algorithms and the typical applications that are associated with PdM. As I have mentioned, I will be looking at ML on a high-level overview for my document; however, I will definitely be discussing in further detail, in separate documents, the sub-field of machine learning and how it helps predictive maintenance and optimizes reliability. A study that was carried out in 2015 (Huang et al., 2015) looked into the application of a machine learning algorithm known as "Support Vector Machine (SVM)" to forecast the amount of time that is still left for certain components to be useful. They argued that because SVM can swiftly handle short training sets and multi-dimensional data, it is a viable technique for predicting RUL as explained by Kang et al. (2012).

Alsina et al. (2018) compared various ML algorithms to more conventional statistical techniques, such as the Weibull distribution, by employing the supervised learning methodology. They looked at four different supervised learning algorithms, such as artificial neural networks, SVM, random forests, and different kinds of soft computing. These results showed that machine learning algorithms are very good at predicting how reliable parts will be. Random forest models have always been the most accurate and led to the best results. They also demonstrated that the use of censored data or pre-processed data to improve performance for all of the algorithms that were taken into consideration was a problem. The performance of ML algorithms was much better than that of more traditional methods like the Weibull distribution (Alsina et al., 2018). According to the findings of a research paper, prognostics and RUL prediction, both of which are essential components of maintenance planning, make use of semi-supervised learning methods (Xu and Saleh, 2021; Macedo et al., 2022; Welte et al., 2020). Unlike unlabeled data, which is plentiful, labeled data (such as failure) is uncommon, expensive, or difficult to collect. However, reliability and safety applications offer enormous potential that, in our opinion, is underutilized.

Xu and Saleh (2021) say that supervised regression is often used to estimate the remaining useful life (RUL) and predict when something will break down. Literature in this field talks about how machine learning can be used to improve different kinds of technical parts. Some of these parts are lithium-ion batteries, rail tracks, tools for cutting turbine blades, rolling bearings, and airplane engines. For the purposes of engineering surrogate modelling, RUL prediction, fire hazard simulation, and more generally for structural reliability challenges, researchers have investigated several machine learning models, such as support vector machines, regression, Gaussian Process Regression (GPR) or kriging, and deep learning. In the process of defect detection and identification, unsupervised classification is used quite frequently. This requires not only determining the different types of breakdowns that can occur but also performing online monitoring of the equipment degradation statuses (binary and multi-class). In this context, classification is at the crossroads of two more general issues, namely PHM and predictive maintenance; it offers vital information that supports both practices, and it sits at the crossroads of both. Clustering, which is also known as unsupervised classification, does not see nearly as much use in the context of reliability and safety applications as the supervised version of the same classification method. It has a vast unfulfilled potential that the aforementioned domains of image recognition, genomics, and e-commerce are more positioned to exploit than is now being done. One of the applications of clustering is the identification of bearing faults in rotating machines. Other applications of clustering include the classification of damage in structural components, the analysis of degradation in railway point machines, the diagnosis of problems with wind turbines, fault detection in the nuclear industry, and fault detection in the aerospace industry. Clustering also has applications in the identification of faults in the nuclear industry.

Xu and Saleh (2021) say that anomaly detection is used a lot to find problems with engineering equipment and structures before they break down. This is because anomaly detection works really well for this kind of application. It has to do with sensor data, and for industrial machinery and equipment, sensor data usually comes in the form of a series of measurements taken over time. In some situations, it is crucial to find problems right away in order to stop them from getting worse and causing a disaster. The data that is used to find problems in things like beams, airframes, and bridges has both a spatial and a temporal dimension. Examples of such structures include bridges, beams, and airframes. Even though we have placed anomaly detection under the category of unsupervised learning, algorithms for anomaly detection do exist in both supervised and semi-supervised learning modes as well. This is despite the fact that we have placed anomaly detection under the category of unsupervised learning. An unsupervised mode is the one that is utilized the majority of the time since it is less expensive (and frequently quicker to obtain), and it has a larger availability of unlabeled data in comparison to labeled data. In applications dealing with dependability and safety, data are usually only accessible under what are known as nominal operational circumstances. Anomaly detection can benefit from semi-supervised learning algorithms in situations when only incomplete labels are given.

Algorithms for semi-supervised learning are used in all important maintenance tasks, such as prognostics, RUL prediction, fault detection and identification, and more. In reliability and safety applications where unlabeled data is abundant but labeled data (such as failure) is difficult to come by, expensive to collect, or difficult to obtain and, in their judgment, has an unrealized potential, they demonstrated a higher RUL prediction accuracy and a lower resulting dispersion as compared to a supervised learning system. Despite this, they used a significantly higher degree of labeled data in their use of semi-supervised learning when compared to the typical percentages found in such cases. Learning that is either supervised or unsupervised is nevertheless applied more frequently in applications dealing with dependability and safety than is learning that is reinforced. Yet, it offers a number of opportunities to make important contributions to these subjects in a variety of ways.

For the purpose of predicting the RUL of components, some systems make use of deep neural networks (also known as deep learning techniques). The results of the experiments that were conducted and published in a publication by Li et al. (2019) indicate that the technique that was presented has potential for solving prognostic concerns and is well suited for use in industrial applications. Even though the suggested method gave rather accurate predicted results, a great deal more labeled training was required in order to successfully train models for neural networks. Techniques that are driven by data typically call for a substantial amount of the underlying data, particularly for predictive algorithms. As can be seen in Figure 2.3, Nicora et al.

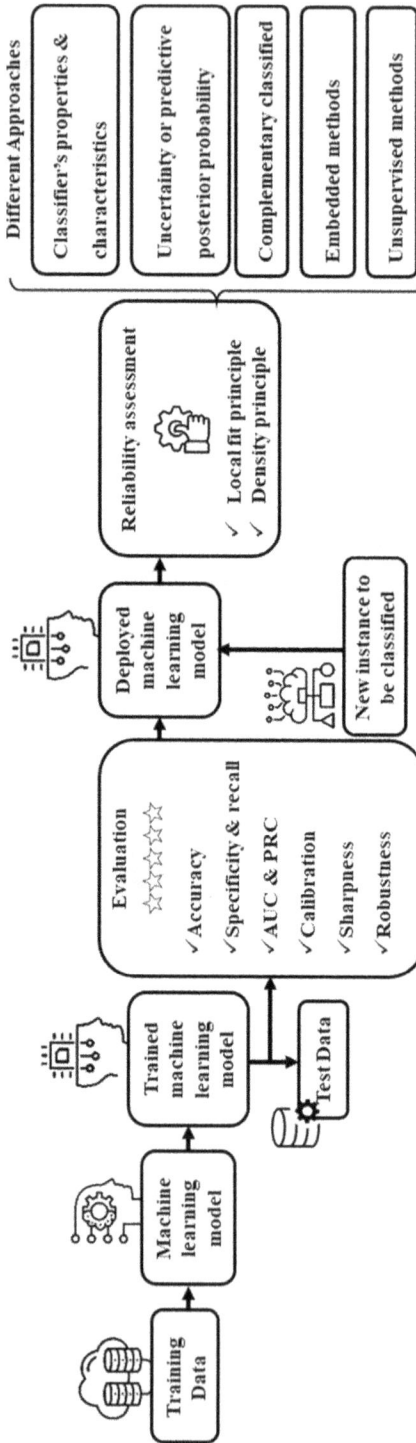

FIGURE 2.3   Machine Learning model life cycle (adapted from Nicora et al., 2022).

(2022) proposed a pathway for the development and deployment of machine learning models. This workflow places an emphasis on validation and continual monitoring through reliability evaluation.

2.4 METHODOLOGY

When previously conducted research is analyzed and reconstructed in a manner that is consistent with its findings, a subject makes intellectual progress (Kumar et al., 2019; Peter et al., 2022). As a result, the purpose of this research is to mine a wide variety of journals and other published publications for articles that include information essential to the application of machine learning in maintenance work and reliability prediction. This could be helpful for both practitioners and academicians in the process of crafting a response that is suited to the requirements of society. When used as a technique of study (Biswal et al., 2018; Snyder, 2019; He et al., 2020; Swain et al., 2021), literature reviews make significant contributions to the conceptual, methodological, and thematic development of a variety of diverse fields (Palmatier et al., 2018; Swain et al., 2022).

During the process of preparing this paper, the primary source of information came from published journals obtained from the databases of "SCOPUS", "Web of Science", "Google Scholar", and "Research Gate". The phrases "Machine Learning," "Deep Learning," "Preventive Maintenance," "Reliability," "Predictive Maintenance," "Artificial Intelligence (AI)," etc., were the most important keywords that were utilized during the search. A total of 50 relevant articles were obtained. Then these were filtered according to our study criteria, and 22 full articles were considered for this study.

2.5 RELIABILITY PREDICTION USING ML

Following the examination of a number of papers, a few remarkable discoveries have emerged. These findings will be highlighted in the following section of the article, and an attempt will be made to derive an implementation strategy that will be addressed in the subsequent sections.

In general, ML has the potential to be a very helpful tool for making predictions about reliability. When organizations use machine learning to predict reliability, they can take preventative steps to avoid equipment failures, reduce downtime, and make their maintenance programs work as well as possible. ML can be used to look at historical data from equipment or systems. This can then be used to predict reliability by finding patterns in the data and predicting future problems. Here's a quick summary of how ML can be used to make predictions about reliability:

- **Data Collection**: The first thing that has to be done is to collect data on the operation and behavior of the system or piece of equipment. Some of the possible ways to get this information are through sensors, monitoring systems, and other sources. The data that are collected can include everything such as temperature, amperage, acceleration, and vibration, etc., but they must have some kind of association with a particular failure mode or modes, or at least a suspected correlation with those failure modes.

- **Cleaning and pre-processing the data**: Once the data has been collected, it needs to be cleaned and pre-processed to get rid of any mistakes or inconsistencies that may have happened while the data was being collected. It's important to remember that data, particularly failure data, can be quite noisy. Failure can be triggered by a number of different modes, and the data collection period typically involves overlap between all of these modes. In the publications referenced above, data preparation is described as cleaning and eliminating excessive nonlinearity in the data. There is a possibility that this non-linearity is unrelated to the component that is now being examined. An example of this would be an operator or technician making a mistake when logging temperature measurements or filling out a Root Cause Failure Analysis (RCFA). Before beginning the training process for machine learning models, these errors will need to be filtered. We want the model to estimate the failure of components, not the random error caused by humans. The data will need to be normalized, missing values will need to be filled in, and outliers will need to be removed.

- **Choosing features**: The next step is to find relevant data features that can be used for prediction. This process is referred to as "feature selection." It is the practice of locating reliable connections within datasets. These attributes exhibit the features of a specific phase in the lifecycle of an entity, and as a result, they are able to be used in the future to identify that stage when comparing them with new data. The feature extraction layer lays the framework for constructing mathematical models that are employed in prediction. A number of elements, such as operating circumstances, environmental considerations, maintenance records, and others, can have an effect on a product's reliability.

- **Training the Model**: Once the important features have been chosen, the ML algorithms are trained on the historical data to find patterns and correlations between the features and the reliability outcomes. The implementation of the ML algorithms begins at this stage, which is the primary area of responsibility. It is up to the authors themselves to determine if the data source contains labeled or unlabeled information. During training, the mathematical model that outlines the connection between the inputs and the outputs is laid out and put through its paces. The ability of machine learning algorithms to infer mathematical relationships from data sets on their own, without the assistance of a human, is one of the many advantages of these algorithms.

- **Model evaluation**: After the model has been trained, it is tested on a separate dataset to determine how accurate and effective it is. The results of this review will help determine whether or not the model could benefit from any modifications.

- **Prediction**: Once the model has been trained and tested, it can be used to make predictions about the reliability of the system or the equipment. This can be accomplished by feeding the model information regarding the current or future operating circumstances, in addition to any other pertinent variables, in order to provide a reliability prediction.

2.5.1 Algorithms Employed

Many algorithms were discussed in the preceding articles as follows:

Type	Algorithm
Unsupervised, Supervised Learning	Support Vector Machines (SVM)
Unsupervised Learning	Deep Neural Networks
Supervised Learning	Random Forests
Supervised Learning	Artificial Neural Networks (ANN)
Supervised Learning	Linear Regression

2.5.2 Advantages of Employing ML in Reliability Prediction

This section outlines some of the benefits that could be gained by adopting ML-based maintenance solutions in organizations.

- **Improved predictive maintenance**: Algorithms that learn from machine data can look at sensor data from machines and equipment to find trends and oddities that could indicate a breakdown is coming. This allows for predictive maintenance to be performed. Maintenance teams are able to foresee possible problems before they arise and take preventative measures to avoid unplanned downtime when they use this strategy because they can predict potential problems before they occur.
- **Better analysis of the root cause**: When anything goes wrong, it is possible to apply machine learning algorithms to figure out what the primary source of the issue is. Machine learning algorithms can find patterns that humans might have missed by looking at data from many different sources, such as equipment logs, maintenance records, sensor data, and so on. This information can be used to make better decisions about how to handle the problem and keep it from happening again in the future.
- **Enhanced asset management**: One can use machine learning algorithms to get the most out of their machines and other equipment. These algorithms can assist businesses in determining which assets are the most important, which require the most upkeep, and which can be replaced or retired by analyzing data on consumption, maintenance, and other criteria. With the help of this information, more informed choices can be made about the distribution of resources and the management of the maintenance of assets.
- **Optimization of the supply chain**: Algorithms that are based on ML can be used to improve the supply chain for operations related to maintenance. These algorithms can assist firms in optimizing their inventory levels, reducing lead times, and improving overall supply chain efficiency by accessing data on inventory levels, lead times, and other parameters. This can help guarantee that maintenance tasks are carried out in an effective and efficient manner.

2.5.3   CHALLENGES ENCOUNTERED BY PNG ORGANIZATIONS IN IMPLEMENTING ML BASED MAINTENANCE SOLUTIONS

PNG is a country in the Pacific region that faces a number of problems when trying to use machine learning-based maintenance solutions in its many industries. Among these difficulties the following are the most important.

- **Limited technical expertise**: The implementation of maintenance solutions that are based on machine learning necessitates the possession of technical competence in the fields of data science, computer programming, and engineering. But because PNG only has a small number of technical experts in these fields, it is hard for businesses to come up with and use machine learning-based maintenance solutions.
- **Limited data availability**: In order to correctly train and test machine learning models, vast volumes of data are required. On the other hand, PNG's many industries frequently lack the complete data sets that are necessary for the construction of efficient machine learning models. Because of this, it could be challenging to create predictive maintenance models that are accurate.
- **High expenses associated with the infrastructure**: The implementation of maintenance solutions based on machine learning frequently demands considerable financial expenditures in infrastructure, such as sensors, data storage systems, and software. These expenditures could be insurmountable for PNG's various industries, which frequently function with limited budgets.
- **Insufficient connectivity to the internet**: In order to properly train and update machine learning models, continual access to relevant data is required. The absence of dependable internet connectivity in many parts of Papua New Guinea, on the other hand, makes it challenging to ensure that these models receive a steady stream of data.
- **Lack of Awareness**: Lastly, many PNG businesses may not know the potential benefits of machine learning-based maintenance solutions or how to use them correctly. This could be due to a lack of awareness. Because of this lack of awareness, it may be difficult for these industries to place a priority on the implementation of these solutions.
- **Resistance to change**: The conventional way of thinking that most employees have, as well as their opposition to change and the adoption of new technologies, is another important barrier that will prevent the successful implementation of ML based maintenance strategy within the local industries of this beautiful country of PNG.

2.6   IMPLICATIONS OF THE RESEARCH

This research suggested several implications, which are listed below:

This research has revealed that one of the biggest factors stopping PNG organizations from using ML-based maintenance practices is the lack of data and ways to collect it. This is particularly true for businesses that do not already conduct reliability assessments or that do not collect data on product failures. It's possible that the mining and oil and gas industries will fare better in the implementation process because they've already put reliability engineering strategies into practice and are collecting reliability data. Because of this lack, there will be an increased cost associated with the installation due to the purchase of equipment for reliability analysis. This revelation could be helpful for organizations to build a system to gather accurate information.

Another thing to think about is the amount of computing power that is needed to train machine learning models. A lot of computer resources are used up when an algorithm for machine learning is being trained. For businesses to do their own processing of machine learning algorithms, they may need to buy graphics processing units (GPUs) and servers. This problem can be solved by utilizing cloud computing systems that are designed for machine learning, such as Amazon Web Services (AWS) or Google Cloud Platform (GCP). It is possible to train models on the cloud, and prognostics may be executed on computers with less processing capability.

The other inhibiting factor identified by this research is shortage of skilled manpower to implement these strategies. Hence, the organizations could partner with government or universities to develop facilities to train human resources.

2.7 RECOMMENDATIONS

To get started, we will address the requirements for such systems. At the moment, the economy of Papua New Guinea has reached a point where it is becoming increasingly expensive to continually import components from other countries. Not only is it expensive, but it also takes a significant amount of time to import goods, which is why inventory management tactics involve stockpiling components and maintaining enormous warehouses for machine parts. The cumulative effect of all of these factors is an increase in the company's operational expenses and a lower profit margin. Just-In-Time (JIT) inventory management strategies can be applied in an organization if it is possible to correctly anticipate the level of an asset's reliability to a given degree of accuracy. Not only will there be a rise in inventory, but there will also be an increase in the availability of assets and the reliability of those assets. This will enable businesses to maximize production and better manage resources such as manpower.

The utilization of machine learning as a tool for the enhancement of maintenance activities in Papua New Guinea can be a highly effective strategy. The following are some suggestions for how to approach the utilization of machine learning to improve maintenance:

- **Determine the difficulties associated with maintenance**: The first stage is to determine the difficulties associated with maintenance in Papua New Guinea. These could involve problems such as broken equipment, concerns about the safety of the workers, or a lack of resources.
- **Implement reliable system to gather data**: After the issues associated with maintenance have been determined, the next step is to collect data in order to train the machine learning models. This might include information on how the equipment is used, logs of its maintenance, and safety records.

- **Determine which machine learning algorithms are most appropriate**: Several different machine learning techniques might be suitable solutions, depending on the nature of the problem with the upkeep. For instance, predictive maintenance could make use of time series analysis or regression models, whereas worries over safety could make use of categorization techniques.
- **Train the machine learning models**: With the right data and algorithms, machine learning models can be taught to forecast equipment failures, identify safety hazards, and optimize maintenance schedules.
- **Incorporate machine learning into maintenance procedures**: The fifth and last phase is to incorporate machine learning models into the procedures that are used for maintaining equipment in Papua New Guinea. This may require generating dashboards or warnings for the maintenance crew, as well as automating the scheduling of maintenance.
- **Skill development of human resources**: It is essential to keep in mind that even if machine learning has the potential to be an effective tool, it is not a panacea. It is still important to have well-trained staff for maintenance procedures to work well, and there are many situations where human judgment is needed.
- **Improvement of internet infrastructure**: A highly reliable and seamless internet connectivity infrastructure would be helpful in successful implementation of ML based maintenance strategies.

It is essential to emphasize the fact that using machine learning to enhance maintenance tasks calls for a large commitment of both time and resources. In addition, it is essential to make certain that any machine learning models are routinely examined and updated to consider variations in the performance of the equipment as well as the characteristics of the environment.

One last thing to take into consideration is how heavily experienced individuals are relied upon. Companies put experienced staff in each region because these workers know the system well enough to predict and stop any breakdowns. If the process is looked at through the lens of machine learning and knowledge engineering, scaling can be built into expert systems. Hence, these experienced individuals will have a higher rate of productivity, and their effective range will be expanded to include numerous locations per individual.

2.8   CONCLUSION

In PNG, where resources and infrastructure are often not as good as they could be, using ML to help with maintenance can be very helpful. When businesses improve the efficiency and effectiveness of their maintenance efforts, they can cut down on downtime, make more money, and make things safer. In addition to this, ML can assist organizations in making more informed decisions regarding the distribution of resources and the prioritization of maintenance activities, both of which can help to ensure that essential pieces of equipment and infrastructure are properly maintained. It is also challenging for PNG's industries to embrace ML-based maintenance solutions due to the limitations of inadequate technical skills, data availability, high infrastructure costs, restricted internet access, and a lack of awareness. But, by solving these issues through investments in technical training, data collecting and storage, infrastructure, and education, PNG's industries will be better able to make use of the benefits that these technologies offer. PNG is in the third world and is just starting to build its IT infrastructure. Also, its economy isn't stable right now, so a company that invests in the ML approach to predict the reliability of its assets will definitely learn something that will change its life. It will save the organization money, cut down on downtime, improve the performance of the assets by making them more reliable, and protect the equipment from random failure by closely monitoring the performance of the asset(s) and fixing it before it fails. All of these benefits will accrue simultaneously. The ML approach can really help in predicting the reliability of assets for any organization so long as there is proper data collection from, say, sensors or PLCs. Because of the money that the organization has been able to save by lowering the cost of downtime, eliminating the cost of replacing unnecessary components before they actually fail, cutting the cost of labor in general, and avoiding accidents, it has been able to contribute more monetarily to the nation by either funding additional activities or improving the standard of living.

2.9 LIMITATIONS AND SCOPE OF FUTURE WORK

The current investigation was done by looking at information that had already been published. However, there was no statistical confirmation of the claims. This is a significant shortcoming of the paper that needs to be addressed. In future research projects, professionals from a variety of industries who are experts in their field will be interviewed about their thoughts on the factors that influence the adoption of ML-based maintenance approaches in PNG and other Asia-Pacific countries. In further research, statistical methods will also be used to investigate the level of influence exerted by the various components.

REFERENCES

Afshari, Sajad Saraygord, Fatemeh Enayatollahi, Xiangyang Xu, and Xihui Liang. "Machine learning-based methods in structural reliability analysis: A review." *Reliability Engineering & System Safety* 219 (2022): 108223.

Alsina, Emanuel F., Manuel Chica, Krzysztof Trawiński, and Alberto Regattieri. "On the use of machine learning methods to predict component reliability from data-driven industrial case studies." *The International Journal of Advanced Manufacturing Technology* 94, no. 5 (2018): 2419–2433.

Biswal, Jitendra Narayan, Kamalakanta Muduli, Suchismita Satapathy, and Sushant Tripathy. "A framework for assessment of SSCM strategies with respect to sustainability performance: an Indian thermal power sector perspective." *International Journal of Procurement Management* 11, no. 4 (2018): 455–471.

Çınar, Zeki Murat, Abubakar Abdussalam Nuhu, Qasim Zeeshan, Orhan Korhan, Mohammed Asmael, and Babak Safaei. "Machine learning in predictive maintenance towards sustainable smart manufacturing in industry 4.0." *Sustainability* 12, no. 19 (2020): 8211.

Di Napoli, Mariano, Francesco Carotenuto, Andrea Cevasco, Pierluigi Confuorto, Diego Di Martire, Marco Firpo, Giacomo Pepe, Emanuele Raso, and Domenico Calcaterra. "Machine learning ensemble modelling as a tool to improve landslide susceptibility mapping reliability." *Landslides* 17, no. 8 (2020): 1897–1914.

Duchesne, Laurine, Efthymios Karangelos, and Louis Wehenkel. "Recent developments in machine learning for energy systems reliability management." *Proceedings of the IEEE* 108, no. 9 (2020): 1656–1676.

Eddarhri, Maria, Jihad Adib, Mustapha Hain, and Abdelaziz Marzak. "Towards predictive maintenance: the case of the aeronautical industry." *Procedia Computer Science* 203 (2022): 769–774.

He, Xuzhen, Haoding Xu, Hassan Sabetamal, and Daichao Sheng. "Machine learning aided stochastic reliability analysis of spatially variable slopes." *Computers and Geotechnics* 126 (2020): 103711.

Huang, Hong-Zhong, Hai-Kun Wang, Yan-Feng Li, Longlong Zhang, and Zhiliang Liu. "Support vector machine ba sed estimation of remaining useful life: current research status and future trends." *Journal of Mechanical Science and Technology* 29, no. 1 (2015): 151–163.

Kang, Jianshe, Xinghui Zhang, Jianmin Zhao, and Duanchao Cao. "Gearbox fault prognosis based on CHMM and SVM." In *2012 International Conference on Quality, Reliability, Risk, Maintenance, and Safety Engineering*, pp. 703–708. IEEE, 2012.

Kumar, Sachin, Prayag Tiwari, and Mikhail Zymbler. "Internet of Things is a revolutionary approach for future technology enhancement: a review." *Journal of Big Data* 6.1 (2019): 1–21.

Lees, Frank. *Lees' Loss prevention in the process industries: Hazard identification, assessment and control.* Butterworth-Heinemann, 2012.

Li, Xiang, Wei Zhang, and Qian Ding. "Deep learning-based remaining useful life estimation of bearings using multi-scale feature extraction." *Reliability Engineering & System Safety* 182 (2019): 208–218.

Macedo, Luísa, Luís Miguel Matos, Paulo Cortez, André Domingues, Guilherme Moreira, and André Luiz Pilastri. "A Machine Learning Approach for Spare Parts Lifetime Estimation." *ICAART* 3, pp. 765–772. 2022.

Nicora, Giovanna, Miguel Rios, Ameen Abu-Hanna, and Riccardo Bellazzi. "Evaluating pointwise reliability of machine learning prediction." *Journal of Biomedical Informatics* 127 (2022): 103996.

Palmatier, Robert W., Mark B. Houston, and John Hulland. "Review articles: Purpose, process, and structure." *Journal of the Academy of Marketing Science* 46 (2018): 1–5.

Paolanti, Marina, Luca Romeo, Andrea Felicetti, Adriano Mancini, Emanuele Frontoni, and Jelena Loncarski. "Machine learning approach for predictive maintenance in industry 4.0." In *2018 14th IEEE/ASME International Conference on Mechatronic and Embedded Systems and Applications (MESA)*, pp. 1–6. IEEE, 2018.

Peter, Oyekola, Suchismita Swain, Kamalakanta Muduli, and Adimuthu Ramasamy. "IoT in Combating COVID-19 Pandemics: Lessons for Developing Countries." *Assessing COVID-19 and Other Pandemics and Epidemics using Computational Modelling and Data Analysis* (2022): 113–131.

Ren, Yali. "Optimizing predictive maintenance with machine learning for reliability improvement." *ASCE-ASME Journal of Risk and Uncertainity in Engineering Systems Part B Mechanical Engineering* 7, no. 3 (2021).

Swain, Suchismita, Oyekola Peter, Ramasamy Adimuthu, and Kamalakanta Muduli. "Blockchain technology for limiting the impact of pandemic: Challenges and prospects." *Computational Modeling and Data Analysis in COVID-19 Research* (2021): 165–186.

Swain, Suchismita, Peter Oluwatosin Oyekola, and Kamalakanta Muduli. "Intelligent Technologies for Excellency in Sustainable Operational Performance in the Healthcare Sector." *International Journal of Social Ecology and Sustainable Development (IJSESD)* 13, no. 5 (2022): 1–16.

Sarker, Iqbal H. "Machine learning: Algorithms, real-world applications and research directions." *SN Computer Science* 2.3 (2021): 160.

Snyder, Hannah. "Literature review as a research methodology: An overview and guidelines." *Journal of Business Research* 104 (2019): 333–339.

Theissler, Andreas, Judith Pérez-Velázquez, Marcel Kettelgerdes, and Gordon Elger. "Predictive maintenance enabled by machine learning: Use cases and challenges in the automotive industry." *Reliability Engineering & System Safety* 215 (2021): 107864.

Welte, Rebecca, Manfred Estler, and Dominik Lucke. "A method for implementation of machine learning solutions for predictive maintenance in small and medium sized enterprises." *Procedia CIRP* 93 (2020): 909–914.

Wu, Leon, Gail Kaiser, David Solomon, Rebecca Winter, Albert Boulanger, and Roger Anderson. "Improving efficiency and reliability of building systems using machine learning and automated online evaluation." In *2012 IEEE Long Island Systems, Applications and Technology Conference (LISAT)*, pp. 1–6. IEEE, 2012.

Xu, Zhaoyi, and Joseph Homer Saleh. "Machine learning for reliability engineering and safety applications: Review of current status and future opportunities." *Reliability Engineering & System Safety* 211 (2021): 107530.

3 Quality Control in the Era of IoT and Automation in the Context of Developing Nations

Kialakun N. Galgal
PNG University of Technology, Lae, Papua New Guinea

Manidatta Ray
Birla Global University, Bhubaneswar, India

Bikash Ranjan Moharana and Bikash Chandra Behera
C. V. Raman Global University, Bhubaneswar, India

Kamalakanta Muduli
PNG University of Technology, Lae, Papua New Guinea

CONTENTS

DOI: 10.1201/9781003346623-3

3.1   INTRODUCTION

Quality control in the era of the Internet of Things (IoT) and automation is a very broad theory of its own. Therefore, specifically for this document, more emphasis will be given to quality control practices using IoT and automation in manufacturing industries. Quality concerning the IoT and automation in itself is the ability to monitor and control the equipment remotely or automatically and improve the inspection and measurement reliability. This eliminates errors caused by human intervention and improves consistency or accuracy along the production line.

For manufacturing companies, IoT aids in predictive maintenance, monitoring of production lines remotely, automating a production process, and others not named. Manufacturing organizations all over the world have seen positive development after the adaptation of IoT and automation in their industries ever since its evolution into the industries. They were able to monitor and track production using data collected by sensors along the production line. The implementation of the IoT devices is very important when it comes to the automation of a process line.

Though there may be setbacks or limitations to this technology for a third-world country like PNG, there is vast room for advancement of this practice that can be adopted by all manufacturing industries in this country. The following Sections 3.1.1 to 3.4 will provide the reader with the background of quality control in the era of IoT and automation-manufacturing/production, the problem statement developed from the research conducted, the objectives directed towards the problem statement, and the proposed solution methodology to minimize or eliminate the problem(s).

3.1.1   Background

The Internet of things refers to connected devices which exchange data over a wired or wireless network depicted in Figure 3.1. The IoT platform is the core of automation in any aspect of an industry. With that understanding, the following information will provide how quality control functions with the assistance of IoT and automation in a manufacturing or production facility. Before getting in-depth into the topic, it is best for the reader to understand how the IoT platform has evolved into how it is today.

Before the internet was established in the late 1980s, the platform called *ARPANET (Advanced Research Project Agency Network)* was used. It was firstly initiated in 1966 by a group of computer scientists, allowing scholars and researchers to exchange information between different locations (Cveticanin, 2022). ARPANET was first used in 1969 and then decommissioned in 1989. Research has shown that ARPANET

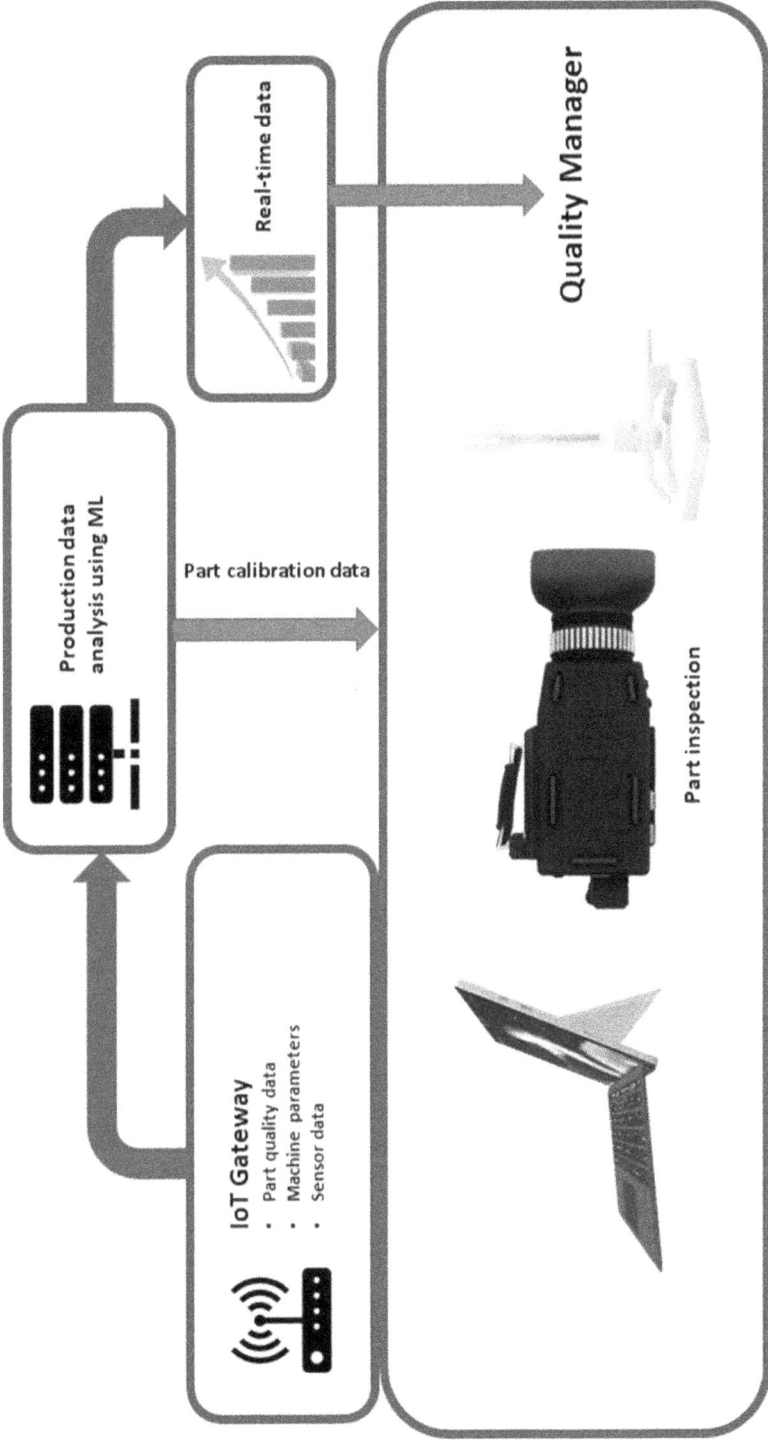

FIGURE 3.1 Quality control using Internet of Things (IoT).

was the first network to have connected two computers successfully without major issues, and all the history of IoT starts with it.

From the early 1990s to the early 2000s, the use of the internet rapidly increased. Industries have understood and realized the use of IoT devices such as sensors, transmitters, controllers, etc., and how these devices store, transmit, and process data to monitor and control the processes and allow timely intervention to be made. More and more manufacturing industries have adopted this practice, which has proven to have helped their organization save costs and make sound decisions. Nowadays, there are more than 30 billion users of IoT devices globally (Hijazin and Zhang, 2019).

The evolution of the Internet of Things had by far resulted in a lot of virtual connections that affected real-life items or activities. Table 3.1 describes the evolution of IoT according to (Tech Ahead, 2022).

Generally, an IoT system comprises three main layers, which are the physical layer (devices, sensors, controller, etc.), the edge computing layer (Storing and Processing of data), and the application layer (Set of integrated services) shown in Figure 3.2.

With the brief background information on the evolution of the Internet from ARPANET provided earlier, it is understood that globally most manufacturing industries have adopted this practice. In terms of quality control, IoT and automation aid in measuring and monitoring equipment so timely human intervention can occur where necessary to safeguard the equipment and men along the production process.

3.1.2 Problem Statement

Quality control can be improved with the use of IoT devices, especially in manufacturing industries here in PNG. Since PNG is a developing nation, the trend of ICT innovation here is much slower than what the developed nations have already

---

TABLE 3.1

The Evolution and Timeline of IoT according to (Tech Ahead, 2022)

Year	Founder	Activity
1982	David Nichols	Coke Vending Machine Project – Carnegie Mellon University
1989	Tim Berners Lee	World Wide Web Framework creation
1990	John Romkey	Toaster connected to the internet
1993	Quentin Stafford-Fraser and Paul Jardetzky	Trojan Room Coffee Pot was built, University of Cambridge
1999	Kevin Ashton	Development of IoT (Named through project presentation) when linking RFID in P&G's supply chain to the Internet.
2005	United Nations International Telecommunications Union	acknowledged the impact of IoT in its report
2008	Zurich	First IoT conference-Switzerland
2011	Gartner	Hype cycle for emerging technologies
2013	IDC	The report predicted IoT growth of 7.9% by 2020

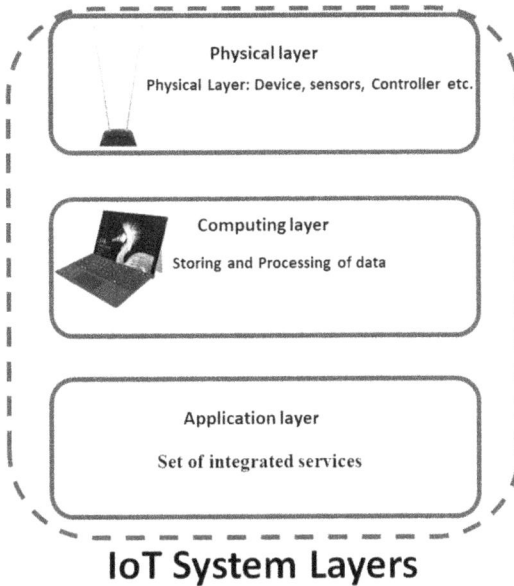

IoT System Layers

FIGURE 3.2   IoT system layers.

utilized. From research conducted, it was identified that IoT devices and applications assist with a variety of industries, be it manufacturing, quality control, logistics, and others. This document however focuses on how quality control can function in this era of IoT and automation, specifically of a manufacturing or production department of an organization. With that being said, we need more awareness and educational or practical guidance towards how IoT can be used to improve quality control processes of our local manufacturing industries operating in PNG.

3.2   OBJECTIVES

The main objectives or aim of this document is to ensure readers, users, or viewers will understand the following:

- Why the use of IoT and automation is important for a manufacturing industry's quality control processes?
- How quality control can be improved in this era of IoT and automation of a manufacturing industry?

3.3   PROPOSED SOLUTION METHODOLOGY

As described in the problem statement and objectives identified from it, research was done on scholarly articles to determine how best this innovation can help improve and address the objectives of this document. Articles similar to the scope of how quality control practices can be practiced in this era of IoT and automation have been reviewed and information gathered will be analyzed and compared to its practicality in a developing nation like PNG. How it can practically improve the quality control processes will be further discussed, and implications will be derived from it. From there, corrective actions will be recommended and will reflect the objectives outlined to conclude this document.

3.4 LITERATURE REVIEW

This section will provide information gathered from previous studies by different authors on how the era of IoT and automation has influenced quality control specifically in the manufacturing or production environment. Even though the articles reviewed may have different objectives, the main idea of how quality control processes work in this era of IoT and automation will be elaborated on to compare how the ideas can be adopted and implemented in our local manufacturing industries in PNG.

In recent years, production facilities have developed drastically in terms of the introduction of innovation and technology. In some countries, this introduction of a smart production technique known as Intelligence Manufacturing (IoT) has proven to have enhanced the productivity of the firm (Lekan et al., 2023) by applying quality monitoring systems. There are quality control tools like a Pareto Chart, Cost and Effect Diagram, etc., to define, measure, analyze, improve, and control the production processes (Shivajee et al., 2019) some of which are believed to be used by local manufacturing facilities in PNG. For instance, two-wheeler automobile manufacturers in India use quality control tools to reduce the cost of production to keep up with the global markets. Their cost reduction exercise has resulted in cleaner production and more sustainable manufacturing processes with hopes of other manufacturing facilities adopting the quality control tools application in their processes. Quality control practices in the era of IoT and automation in manufacturing industries have improved the production systems and boosted the quality of their products by using instant or real-time quality control systems (Sader et al., 2019). With its positive impact on industries, the economic level of the country will also be improved. This new era, known as industry 4.0, has elevated the production level so that machines can intercommunicate to achieve high-efficiency production systems as well as increasing its productivity. Extending the application outside of production, the industry 4.0 era can be applied to logistics and supply chain business functions of the manufacturing industry as well.

In manufacturing industries, IoT has advanced machine controls using smart devices in this industry 4.0 era. For industrial IoT to be a success, data collection and storage are vital (Haghnegahdar et al., 2022). More and more industries are shifting towards cloud-based manufacturing techniques where monitoring and controlling are done remotely. The industry 4.0 advancement is adopted and used mostly by well-developed nations globally. With the advancement in IoT and automation gaining attention from many industries, more industrial practices about this advancement are being adopted and used that include the quality control processes in the industries (Chowdhury and Raut, 2019). Industries have been able to improve most business functionalities like cost saving, scalability, line efficiencies, connectivity, etc.

A study (Madugula, 2021), has shown that IoT devices used along a manufacturing process line have proven to have provided a lot of detailed information on the processes. The improvement and advancement made the industries smarter, more efficient, and more productive. With the main aim of almost all industries being cost saving, the adaptation of the IoT and automation practices about quality control can help with the timely human intervention of machines and processes along the production lines and surely improve the quality of the product and processes. Looking at all the advancement and improvement activities that have occurred globally, especially in developed nations in this era of IoT and automation, we can do so with our local industries in PNG. This era of IoT can also improve other areas of businesses by driving out existing inefficiencies.

Though IoT is shown to have improved the quality of communications across the different sections of the industry, end users will not be able to use it if it is not well designed or has complicated models. Regardless of this downside, if IoT is successfully installed in an industry, it will increase the efficiency of the processes, improve the quality of products, and industrial equipment, and as a result, bring forth more flexible production (Yildiz, 2021). In line with the Industry 4.0 era, Quality control points can be measured along the production line(s) instead of using the old methodology processes (Godina and Matias, 2019). More studies were being carried out on this to determine how imperative the effect of industry 4.0 on quality control can be. In an industry, if smart sensors were used on a production line but the data collected from it wasn't used to improve the process, it would be considered to not add any value to the operation; but on the other hand, engineers would be able to make sound decisions by doing a root cause analysis to identify where the failure mode is, with an attempt to eliminate it.

Quality control is a very important section, especially in manufacturing industries, since one of their main company objectives is to reduce reject rates and save cost. Therefore, with the introduction of IoT into the industries, management has figured that it helps them stay competitive among all other industries (Burkhalter, 2020). IoT applications and devices help manufacturing industries with predictive maintenance, controlling production processes remotely, and automating production processes. It helps industry with process development and innovation. With IoT systems in place, plant managers can make sound decisions to achieve departmental objectives or performance indicators. Studies have shown that more manufacturing industries are focusing on the integration of IoT and automation to improve the quality control unit of the organization.

3.5 METHODOLOGY

Since this document is pure research, information sources are cited and references are attached. With the focus being on how quality control can be improved in this era of IoT and automation, it is a broad narrative; hence, it was further narrowed to the quality control process of manufacturing industries in PNG.

Even though the advancement in IoT and automation is better in the developed nations, there is still room for improvement in the way quality control processes are being practiced in manufacturing industries in PNG. Manufacturing industries in PNG have the opportunity and potential to adopt this new growth and development of quality control processes in this new industry 4.0 era, specifically in IoT and automation. Studies have been conducted on online scholarly articles and compared to the current local quality control practices of local manufacturing industries. Similarities and differences on how the process operates were identified and will be further analyzed and discussed. With the two main objectives of this document being to understand why it is important for our local industries to adapt this new trend that most industries around the world are already enjoying, and how IoT and automation processes can help improve quality control processes, users can find this document useful if they are in doubt of the benefits of this innovation.

3.6   DATA ANALYSIS

Under this section, information gathered from the articles online will now be analyzed. The findings on how quality control functions in this era of IoT and automation will be presented and detailed determination will be made on the accuracy, relevance, or reliability of the information found.

3.7   FINDINGS AND OBSERVATIONS

According to (Rupareliya, 2022), here is the tabulation of findings gathered (Table 3.2):

TABLE 3.2
How Quality Control Can Help Managers

Findings	Description
Detecting anomalies in workflow	IoT and automation can help quality control managers identify non-conforming quality conditions and flag the issue to the engineering team for assistance.
Adjust environmental changes for quality production	Quality control managers can improve their models with IoT and automation. Existing quality control processes can be optimized.
Monitoring post-production storage conditions	Manufactured goods have a shelf-life. With IoT and automation in place, this issue can be monitored, observed, and conditions identified that can affect the shelf-life of the products. Hence, the new technology can come in handy to the quality control managers.
Remote monitoring	Improves production losses and unnecessary downtime
Condition tracking and predictive maintenance	With the use of sensors, data is generated in real-time, this assists the quality control managers in proactively conducting repairs and avoiding downtime.

Additionally, it was identified that IoT or automation can determine how to identify problems quickly, understand the conditions of a functional failure, use of data transmitted through the sensors to make sound and informed decisions, or track a production plant (Newton, 2021). All in all, IoT was found to have enhanced production in most global manufacturing industries. For instance, in local manufacturing industries here in Lae, they mostly use SCADA or PLC programs to monitor production processes with sensors installed along the line, which helps production personnel, maintenance personnel, and quality control personnel combine and make sound decision where necessary regarding a process failure or identify accurately where the problem is so necessary actions may take place. Quality is usually everyone's responsibility, hence with the help of IoT devices and automation, everyone can identify issues and flag them in time for corrective, preventive, and ultimately predictive maintenance to occur and reduce or avoid downtime.

IoT enables manufacturing industries identify and understand issues, perform corrective actions, and make informed decisions, improve quality control, and increase the product quality (Miller, 2021). Automation makes it easier by having sensors and measuring devices placed on critical spots along the production line. An author from IBM (Bigos, 2017), discussed how quality control was improved from being reactive to proactive with the usage of IoT devices and applications. IBM identified that, prior to the introduction of IoT in quality control of manufacturing processes, a lot of guess work was done when there was a component failure relating to low quality products or machine quality. After IoT was introduced and used, improvement was seen in the processes as identified through an increase in quality products or an improvement in downtime loss (Figure 3.3).

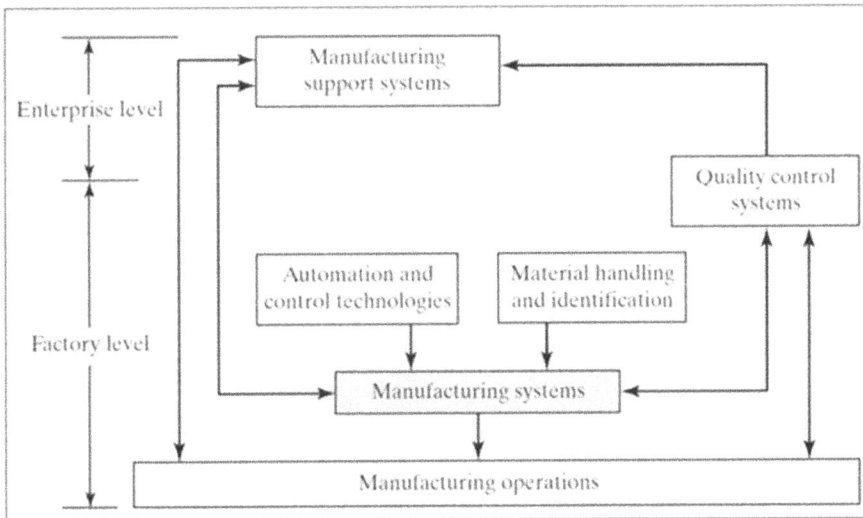

FIGURE 3.3 Relationship Between IoT/Automation, and Quality Control in Manufacturing.

3.8   DISCUSSION

Implementation of IoT and automation processes to improve quality control processes in this industry 4.0 era comes with its advantages, disadvantages, and its importance. The subsections below will provide examples of the advantages, disadvantages, and the importance.

3.9   ADVANTAGES AND DISADVANTAGES OF IoT AND AUTOMATION IN MANUFACTURING

Tabulated here are some advantages and disadvantages of IoT and automation in manufacturing processes. Though it is generic, it is also applicable to the quality control process as it is one of the important processes in a manufacturing industry (Table 3.3).

3.10   IMPORTANCE OF ADOPTING QC PROCESSES IN THIS ERA OF IoT AND AUTOMATION

In light of the IoT and Automation era it is paramount that quality control processes be automated, as it:

- Improves inspection measurement and reliability
- Increases customer loyalty
- Improves quality inspection
- Reduces mistakes and wastages
- Improves efficiency
- Creates an avenue for more participation from all employees
- Customer needs can be better understood
- Reduces operating cost
- Increases performance

TABLE 3.3
Advantages and Disadvantages of IoT and Automation in Manufacturing

Advantages	Disadvantages
Lead time processing	Data security
Communication	Privacy
Monitoring and control	Complexity
Saves time and effort	Compliance
Cost savings	Expensive to implement
Real time tracking	Creates unemployment
Improve product and machine quality	

3.11 IMPLICATIONS

With a better understanding of how quality control processes function in the era of industry 4.0, this document can help users understand why and how it can be helpful for manufacturing industries in PNG to adopt the current global practices of quality control.

3.12 RECOMMENDATION

Since PNG is a developing nation, this IoT and automation will only be understood by those trained on its functionality, purposes, benefits, advantages, and disadvantages of this innovation in terms of quality control processes in a manufacturing industry. Not everyone in this nation knows how to access information from the internet because of our literacy rate. Therefore, in order for this concept to be understood and accepted by end users, here are a few recommendations on how to make this possible:

- Create more awareness about this innovation.
- Disclose benefits so everyone can understand.
- Provide detailed training for this practices prior to implementation.

3.13 CONCLUSION

In understanding why the use of IoT and automation is vital for the quality control process of a manufacturing industry, consider that these align with the main objective of most industries, reduction of downtime and cost savings. This new innovation is very practical and beneficial if it is applied and used well. Additionally, though this will be an expensive investment for the industry, if sourced from renowned software industries like Microsoft, Google, IBM, etc., the company's information, data security, and privacy is guaranteed. Moreover, the industry's quality control needs can be fully met and an improvement in the processes and systems can be experienced.

3.14 LIMITATION AND SCOPE OF FUTURE WORK

The main limitation to this trend will be the educational purposes of end users. They have to be trained to understand the use and importance of how quality control can function in this era. More studies should be carried out on how the concept of IoT and automation can be fully rolled out to a developing nation so that its benefits and importance to the manufacturing industries can boost their performance, productivity, save cost, and minimize loss.

REFERENCES

Bigos, D. (2017, March 08). *From reactive to proactive quality management with IoT*. Retrieved from IBM: https://www.ibm.com/blogs/internet-of-things/quality-management-iot/
Burkhalter, M. (2020, August 20). *Quality control in the era of IoT and automation*. Retrieved from Perle: https://www.perle.com/articles/quality-control-in-the-era-of-iot-and-automation-40189762.shtml#:~:text=While%20quality%20control%20is%20a,that%20waste%20time%20and%20money

Chowdhury, A., and Raut, S. A. (2019). Benefits, challenges, and opportunities in adoption of Industrial IoT. *International Journal of Computational Intelligence & IoT* 2(4), 1–7.

Cveticanin, N. (2022, August 19). *ARPANET: The Project That Launched the Internet.* Retrieved from Data Prot: https://dataprot.net/articles/what-is-arpanet/

Godina, R., and Matias, J. C. (2019). Quality Control in the Context of Industry 4.0. In *Industrial Engineering and Operations Management II: XXIV IJCIEOM*, Lisbon, Portugal, Springer International Publishing (pp. 177–187). Springer.

Haghnegahdar, L., Joshi, S. S., and Dahotre, N.B. (2022). From IoT-based cloud manufacturing approach to intelligent additive manufacturing: industrial Internet of Things—an overview. *The International Journal of Advanced Manufacturing Technology*, 1–18.

Hijazin, K., and Zhang, T. (2019). The application of IoT technology to a manufacturing process: Case study. In *2019 international conference on quality, reliability, risk, maintenance, and safety engineering (QR2MSE), IEEE*, 203–210.

Lekan, A., Aigbavboa, C. and Emetere, M. (2023). Managing quality control systems in intelligence production and manufacturing in contemporary time. *International Journal of Construction Management* 23(8), 1436–1446.

Madugula, L. (2021). Applications of IoT in Manufacturing: Issues and Challenges. *Journal of Advanced Research in Embedded System*, 8, 3–7.

Miller, O. (2021, August 4). *Tips On Enhancing Product Quality in the Industrial IoT Era.* Retrieved from Industrial IoT: https://4iplatform.com/blog/tips-on-enhancing-product-quality-in-the-industrial-iot-era/

Newton, E. (2021, November 23). *IoT Strategies to Optimize Quality Control in Manufacturing.* Retrieved from IoT Times: https://iot.eetimes.com/iot-strategies-to-optimize-quality-control-in-manufacturing/

Rupareliya, K. (2022, January 11). *Quality Control In The Age Of IoT Development Services: A Quick Glance.* Retrieved from Intuz: https://www.intuz.com/blog/quality-control-in-the-age-of-iot-development-services

Sader, S., Husti, I., and Daroczi, M. (2019). Quality Management Practices in the Era of Industry 4.0. *Zeszyty Naukowe Politechniki Częstochowskiej Research Reviews of Czestochowa University of Technology* 35(1), 117–126.

Shivajee, V., Singh, R. K., and Rastogi, S. (2019). Manufacturing conversion cost reduction using quality control tools and digitization of real-time data. *Journal of Cleaner Production*, 237, 117678.

Tech Ahead. (2022). *Evolution of Internet of Things (IoT): Past, present and future.* Retrieved from Tech Ahead: https://www.techaheadcorp.com/knowledge-center/evolution-of-iot/

Yildiz, A. (2021). Use of intelligent manufacturing systems in industries in the era of industry. *Journal of Engineering Research and Applied Science*, 10, no. 2, 1875–1880.

4 Precision Positioning of Robotic Manipulators in Manufacturing Processes through *PID* Controller to Contribute towards Sustainability

Shiv Manjaree Gopaliya
VIT Bhopal University, Sehore, India

Manoj Kumar Gopaliya
The NorthCap University, Gurugram, Haryana, India

CONTENTS

4.1 INTRODUCTION

With the ever-increasing emphasis on the conservation of natural resources, the need for sustainable growth and development in every aspect of life is becoming the necessity of the hour. This is equally true for industrial processes as well, where the concept of conserving energy is becoming paramount. With the upgrading of manufacturing industries from Industry 4.0 to higher, a lot of focus is being paid to reducing energy conservation; thus making it sustainable. In present times, robots are the best and most technically advanced machines available to the manufacturing world. Any industrial application, like welding, pick and place, spray painting, etc., mainly depends on two factors, viz quality and quantity, to fulfill the demands of the highly uncertain market in the present times. Robots play a major part in maintaining

DOI: 10.1201/9781003346623-4

this supply and demand chain of manufacturing industries. However, due to high energy consumption, robots are making these processes unsustainable.

Efforts are made by researchers across the globe to find means and ways to make robots more compatible with the requirements of sustainable growth. Bugmann et al. (2011) have proposed methods to attain sustainable solutions through new work concepts, new component designs, or the use of new materials for industrial robots. Ogbemhe et al. (2017) have highlighted the adoption of Robotic Operating System (ROS) technology in the design aspects of an industrial robot to achieve sustainability at a low cost. While the young researchers are trying to get long-term sustainable solutions with the help of robots, Park (2016) and Dias (2005) have emphasized the role of education in robot-based sustainable development.

Since the quality and finesse of the final product developed using a robot mainly depends on the proper positioning of the end-effector of an industrial robot, many efforts have been made to improve the performance of the end-effector. Nyein et al. (2019) have proposed a fuzzy-based *PID* controller for position control of a 2-DOF planar robot. Nayak and Singh (2015) have proposed a computationally efficient approach for tuning *PID* controllers. The *PID* controller is designed to control velocity as well as path planning of mobile robots (Nguyen et al., 2018; Sokunphal et al., 2018). Coronel-Escamilla et al. (2018) have highlighted the effective use of fractional PI and PD controllers for trajectory planning of SCARA (Selective Compliant Assembly Robot Arm) industrial robots.

Some specific attempts have also been made to achieve sustainable solutions for particular robot applications. For example, Jaafar et al. (2014) have developed and tuned *PID* controllers to control liquid levels for coupled tank systems used in chemical industries. Wu et al. (2018) have proposed efficient and sustainable solutions for power systems. Experimental verification to improve the control performance of a boiler system has been presented using the proposed method. Fiducioso et al. (2019) have proposed an artificial intelligence-based self-tuning PI controller to achieve sustainable solutions for heating, ventilation, and air conditioning units. Qin et al. (2019) have provided simulated results for artificial intelligence based *PID* controller tuning for a fuel cell and fuel utilization, with the least disturbance and regulated output voltage. Parvin et al. (2021) have reviewed conventional methods and intelligent methods to achieve sustainable solutions for optimized building energy management in future directions. An optimized *PID* controller has been designed and implemented on an adaptive cruise control system to improve its performance (Chaturvedi and Kumar, 2021).

During the current year, Zafar et al. (2022) have investigated a machine learning approach for a maximum power point tracking technique to harvest maximum power from a thermoelectric generator under all different working conditions. It has been shown experimentally that sustainable solutions for green energy can be achieved in the future with the proposed method. Similarly, Nonoyama et al. (2022) have attempted to optimize energy for optimal motion planning of a pick-and-place industrial robot by tuning *PID* Controller.

Through these efforts, the reduction in energy consumption, as well as operational and commissioning costs, is practically demonstrated. However, constant efforts in this direction are still needed to achieve the goals of sustainability and a circular economy.

With the beginning of the Industry 5.0 revolution and the recent fight against the Corona pandemic, it has become even more important to accelerate the efforts towards achieving sustainability in every aspect of industrial processes.

The work presented in this chapter highlights the importance of controlling the movement of the industrial applications-based robot and proposes a new method to achieve the most optimized path for the end-effector of industrial robots to achieve sustainability.

4.2 SUSTAINABLE SOLUTIONS FOR INDUSTRIAL ROBOTS

Today's manufacturing world is ruled by robots. While industrial applications like welding, palletizing, spray painting, assembly, and so on are completed with the help of industrial robots, robots are still believed to be unsustainable machines. Firstly, robots consume lots of energy for proper working. Secondly, it is often argued that robots are the cause of unemployability. Recently, new approaches to finding sustainable solutions for robots are being researched in the forms of (a) energy-efficient robot-based manufacturing processes and (b) human-machine interactive manufacturing environments. In other words, Industry 5.0 is paving the way for industrial robots with sustainability in the time to come.

For any industrial application, proper movement of each link and joint of a robot is of utmost importance in reducing its energy consumption. Proper positioning of robot end-effectors to defined coordinates, reduced jerks in joint movements, reduced friction, and optimized trajectory in a workspace are some of the measures through which the set goal can be achieved.

An easy, simple, and effective *PID* controller design helps in proper control of the movement of industrial robots. During this study, a *PID* controller is used for achieving precision in reaching the desired end-effector position and the controlled velocity of each joint. This work is an attempt to highlight the importance of *PID* controller elements in tuning the most optimized trajectory of the end-effector of a 3-*DOF* industrial robot under study to achieve sustainability goals.

4.3 MATHEMATICAL MODEL OF *PID* CONTROLLER

The present industrial world makes wide use of PID Controllers for proper control and functioning of machines. Its widespread use depends on its quality of robust and efficient control of multiple processes and industrial systems. The other good feature of the *PID* Controller is its simple design and tuning methods. *PID* Controller has three important elements viz. proportional term (P), the integral term (I), and derivative term (D), respectively. Proportional (P) term provides information about the immediate errors present or immediate response of the system. The Integral (I) term provides information about all the previous errors dominant in the system. Derivative (D) term provides prediction about future errors to occur in the system. The gains of each element are given as K_p for the P element, K_i for the I element, and K_d for the

D element of the *PID* Controller. For a better understanding of second-order system response, it is very important to analyze the effects of four major parameters, viz rise time, settling time, peak overshoot, and steady-state error. While the time needed to reach a final value of response is known as rise time, settling time is the time needed for a response to get steady. The steady-state response is the difference between desired output value and the actual output value. Peak overshoot is the difference of magnitude values at peak response and steady-state response. The tuning of each *PID* Controller element results in finding the proper response of the dynamic system. The general characteristics of the three elements of the *PID* Controller individually are given in Table 4.1.

The 3-*DOF* industrial robot in this work represents a second-order dynamic system. The transfer function of the 3-*DOF* industrial robot is given by equation (4.1).

$$\frac{\theta_m}{\theta_d} = \frac{K_p\left(\dfrac{1}{Js+B}\right)\left(\dfrac{1}{s}\right)}{1+\left(K_p+K_ds+\dfrac{K_i}{s}\right)\left(\dfrac{1}{Js+B}\right)\left(\dfrac{1}{s}\right)} \tag{4.1}$$

Figure 4.1 shows the simplified block diagram for the closed-loop *PID* controller, where K_p, K_i, and K_d are the proportional gain, integral gain, and derivative gain

TABLE 4.1
Characteristics of Each Element of *PID* Controller

Response Element	Rise Time	Peak Overshoot	Settling Time	Steady State Error
K_p	Reduce	Rise	Small variation	Reduce
K_d	Small variation	Reduce	Reduce	No variation
K_i	Reduce	Rise	Rise	Eliminate

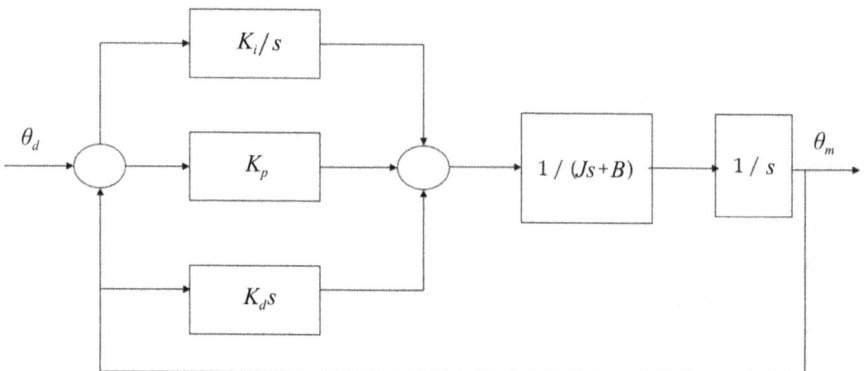

FIGURE 4.1   Schematic diagram of PID controller for 3-DOF industrial robot.

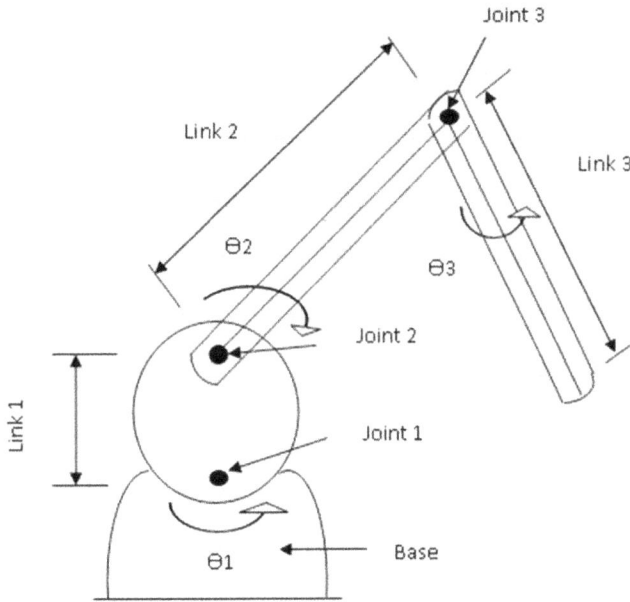

FIGURE 4.2 Schematic representation of 3-DOF Omni-Bundle robotic manipulator (Manjaree et al., 2016).

elements of the controller. θ_m and θ_d are the actual and desired positions of the joints of a 3-*DOF* industrial robot. J is the inertia of the joint and s is the Laplace transform while B is the joint friction.

This work was carried out on a 3-*DOF* industrial robot where the end-effector position is controlled through a PID Controller. The design, specifications, and working of the 3-*DOF* industrial robot can be referred to Manjaree et al. (2016).

Figure 4.2 shows the 3-*DOF* robot used during the current study. The position of joint 1 is defined by $\theta_2 = -\pi/4$ rad, $\theta_3 = \pi$ rad, $J = 0.0031$ kgm^2 and $B = 0.0089$ Nms/rad. The position of joint 2 is defined by $\theta_1 = 0$, $\theta_3 = \pi$ rad, $J = 0.0022$ kgm^2, $B = 0.0170$ Nms/rad. Similarly, the position of joint 3 is defined by $\theta_1 = 0$, $\theta_2 = -\pi/4$ rad, $J = 0.0009$ kgm^2, $B = 0.0058$ Nms/rad.

The 3-*DOF* industrial robot under study consists of three actuated joints and two links with $l_1 = l_2 = 0.132$ mm, respectively.

4.4 EXPERIMENTAL RESULTS AND DISCUSSIONS

The analysis is performed by applying *PID* control to the three actuated joints of the 3-*DOF* industrial robot. Figure 4.3 shows the schematic representation of PID control for the 3-DOF industrial robot under study.

Figure 4.4 shows a triangle trajectory plotted using a 3-*DOF* industrial robot under study.

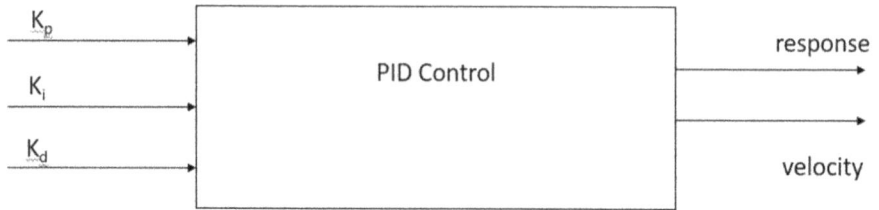

K_p

K_i

K_d

PID Control

response

velocity

FIGURE 4.3   Schematic representation of PID control of 3-DOF industrial robot.

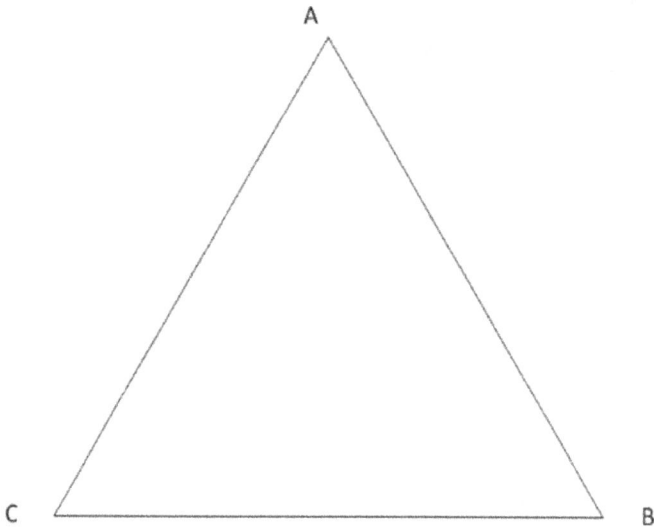

A

C

B

FIGURE 4.4   Plotted trajectory.

To get desired output response of a dynamic system, all the *PID* Controller elements are tuned together. Gains K_p, K_i, and K_d are tuned for the three actuated joints of a 3-*DOF* industrial robot. The effects of tuning each *PID* element can be seen in Figures 4.5a and b, 4.6a and b, and 4.7a and b, respectively.

Figure 4.5a shows the *PID* response for joint 1 of the 3-*DOF* industrial robot under study. After multiple attempts of tuning, the values of gains are achieved as K_p = K_i = 1 and K_d = 1.7. Figure 4.6a shows the *PID* response for joint 2 of the 3-*DOF* industrial robot where the gain values are K_p = 0.6, K_i = 1, and K_d = 1.5, respectively. Figure 4.7a shows the *PID* response of joint 3 of a 3-*DOF* industrial robot with gain values as K_p = 0.4, K_i = 1.1, and K_d = 1.3. Figures 4.5b to 4.7b plot the velocity of each joint of a 3-*DOF* industrial robot.

It is seen from the plots that while increasing the values of K_p accelerates the response of the robot, however, it also lowers the rise time and gives high overshoot values. A better response is received by decreasing the K_p gain value, which in turn decreases the overshoot value and increases the rise time. Similarly, growth in derivative gain K_d values decelerates the response of the robot. However, this step also increases the rise time. In this case, overshoot (introduced due to K_p) is reduced by decreasing the value of K_d.

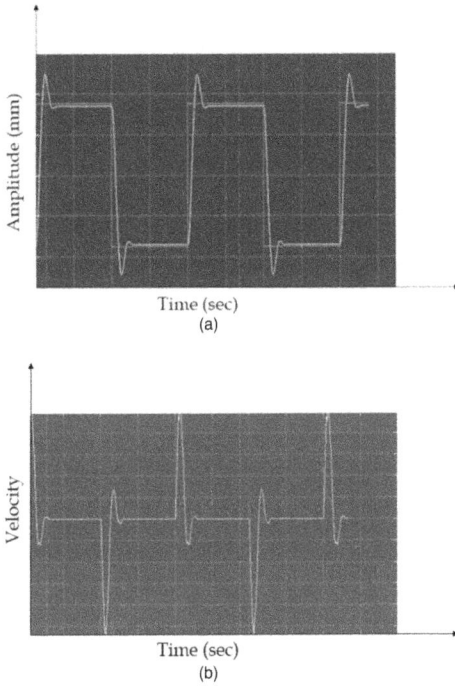

FIGURE 4.5   (a) PID response of joint 1 of 3-DOF industrial robot, (b) Velocity of joint 1of 3-DOF industrial robot.

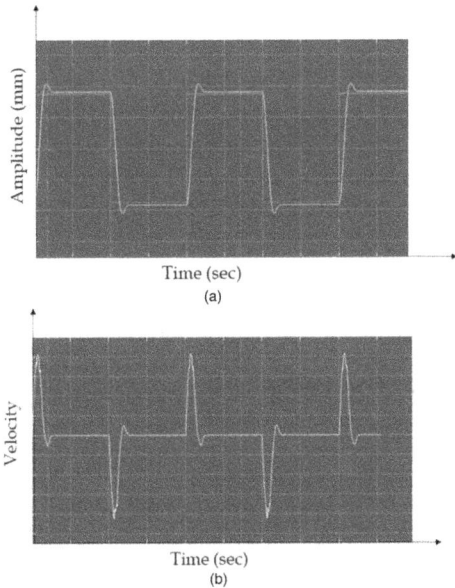

FIGURE 4.6   (a) PID response of joint 2 of 3-DOF industrial robot, (b) Velocity of joint 2 of 3-DOF industrial robot.

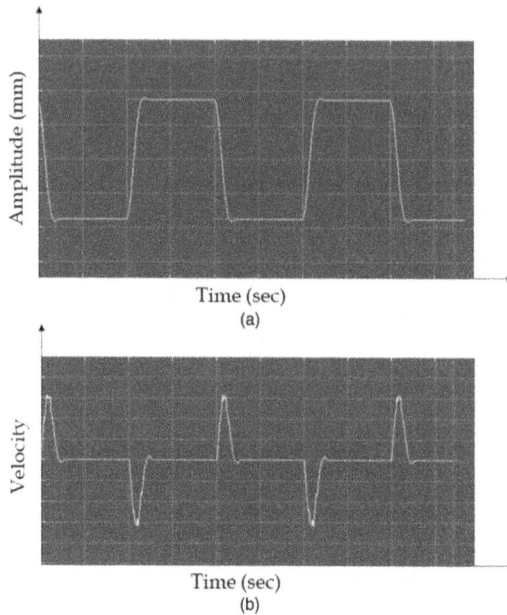

FIGURE 4.7   (a) PID response of joint 3 of 3-DOF industrial robot, (b) Velocity of joint 3 of 3-DOF industrial robot.

4.5   CONCLUSIONS AND FUTURE SCOPE OF THE WORK

Sustainability is the major focus in today's pandemic situation. This work suggests that industrial robots can also be treated as sustainable machines. The concept is tested by achieving precise positioning of robot end-effector for the desired trajectory, using a *PID* controller. Results show that individual tuning of gain of each *PID* element is helpful in optimal response of movement of a 3-*DOF* industrial robot.

A lot of research is needed to find sustainable solutions in all aspects of the use of industrial robots. Further improvement in results is possible with the application of artificial intelligence based on particular industrial applications.

REFERENCES

Bugmann, G., Siegel, M. & Burcin, R. 2011. A role for robotics in sustainable development? *IEEE Africon* 11: 1–4. DOI: 10.1109/AFRCON.2011.6072154.

Chaturvedi, S. & Kumar, N. 2021. Design and implementation of an optimized *PID* controller for the adaptive cruise control system. *IETE Journal of Research*. DOI: 10.1080/03772063.2021.2012282.

Coronel-Escamilla, A., Torres, F., Gómez-Aguilar, et al. 2018. On the trajectory tracking control for an SCARA robot manipulator in a fractional model driven by induction motors with PSO tuning. *Journal of Multibody System Dynamics* 43 (3): 257–277. DOI: 10.1007/s11044-017-9586-3.

Dias, M. B., Mills-Tettey, G. A. & Nanayakkara, T. 2005. Robotics, Education, and Sustainable Development' *Proceedings of the 2005 IEEE International Conference on Robotics and Automation*. 4248–4253. DOI: 10.1109/ROBOT.2005.1570773.

Fiducioso, M., Curi, S., Schumacher, B., et al. 2019. Safe contextual bayesian optimization for sustainable room temperature *PID* control tuning, *Proceedings of the Twenty-Eighth International Joint Conference on Artificial Intelligence*. 5850–5856.

Jaafar, H. I., Hussien, S. Y. S., Selamat, et al. 2014. Development of *PID* controller for controlling desired level of coupled tank system. *International Journal of Innovative Technology and Exploring Engineering* 3 (9): 32–36.

Manjaree, S., Nakra, B. C., Agarwal, V. 2016. Inverse kinematics of 3-*DOF* robotic manipulator using analytical method, ANFIS method and experiments. *International Journal of Mechanisms and Robotic Systems* 3 (4): 297–316.

Nayak, A. & Singh, M. 2015. Study of tuning of PID controller by using particle swarm optimization, *International Journal of Advanced Engineering Research and Studies* 4: 346–50.

Nguyen, Q. M., Tran, L. N. M. & Phung, T. C. 2018. A study on building optimal path planning algorithms for mobile robot. *4th International Conference on Green Technology and Sustainable Development (GTSD)*. 341–46.

Nonoyama, K., Liu, Z., Fujiwara, T., et al. 2022. Energy-efficient robot configuration and motion planning using genetic algorithm and particle swarm optimization. *Energies* 15: 1–20.

Nyein, T., Oo, Z. M. & Hlaing, H. T. 2019. Fuzzy based control of two links robotic manipulator. *International Journal of Scientific Engineering and Technology Research* 8: 1–7.

Ogbemhe, J., Mpofu, K. & Tlale, N. S. 2017. Achieving sustainability in manufacturing using robotic methodologies. *Procedia Manufacturing* 8: 440–46. DOI: 10.1016/j.promfg.2017.02.056.

Park, I. W. & Han, J. 2016. Teachers' views on the use of robots and cloud services in education for sustainable development. *Journal of Cluster Computing* 19: 987–99. DOI: 10.1007/s10586-016-0558-9.

Parvin, K., Hossain Lipu, M. S., Hannan, et al. 2021. Intelligent controllers and optimization algorithms for building energy management towards achieving sustainable development: challenges and prospects. *IEEE Access* 9: 41577–41602. DOI: 10.1109/ACCESS.2021.3065087.

Qin, Y., Zhao, G., Hua, Q., et al. 2019. Multiobjective Genetic algorithm-based optimization of *PID* Controller parameters for fuel cell voltage and fuel utilization. *Sustainability* 11 (12): 1–20.

Sokunphal, T., Othman, W. A. F. W., Alhady, S. S. N., Rahiman, W. 2018. PI controller design for velocity control of mobile robot. *Journal of Fundamental and Applied Sciences* 10: 890–902. DOI: 10.4314/jfas.v10i3s.77.

Wu, Z., He, T., Sun, L., et al. 2018. The facilitation of a sustainable power system: a practice from data-driven enhanced boiler control. *Sustainability* 10: 1–21.

Zafar, M. H., Khan, N. M., Mansoor, M. et al. 2022. Towards green energy for sustainable development: Machine learning based MPPT technique for thermoelectric generator. *Journal of Cleaner Production* 351: 1–17.

5 Role of Additive Manufacturing in Industry 4.0

Harsh Soni and B. N. Sahoo

Sardar Vallabhbhai National Institute of Technology, Surat, India

CONTENTS

5.1 INTRODUCTION

The manufacturing sector is evidence of the massive industrial revolution in the last two decades. The fourth industrial revolution, often known as Industry 4.0, is now taking place as cutting-edge technology becomes more sophisticated every day. Growing automation, intelligent equipment, and industries that use data-driven decision-making to manufacture products more effectively and productively than ever are characteristics of this most recent manufacturing era. Reviewing the past will help us understand where the industry is today. The three previous industrial revolutions laid the groundwork for the industrial technology of today.

5.1.1 FIRST INDUSTRIAL REVOLUTION

With the advent of the first industrial revolution in the late 18th century, it became feasible to create vast amounts of commodities using steam and waterpower as opposed to just human and animal labor. Goods were manufactured utilizing equipment as opposed to being created by hand.

5.1.2 SECOND INDUSTRIAL REVOLUTION

Assembly lines, oil, gas, and electric power were introduced during the second industrial revolution, which occurred a century later. These new power sources enabled mass production and the start of manufacturing process automation.

5.1.3 THIRD INDUSTRIAL REVOLUTION

Beginning in the middle of the 20th century, the third industrial revolution improved production methods by introducing computers, contemporary telecommunications, and data analysis. PLCs, which are used in factories to automate tasks, collect data, and convey information, contributed to the digitization of factories.

5.1.4 FOURTH INDUSTRIAL REVOLUTION (INDUSTRY 4.0)

Industry 4.0 is translating how organizations manufacture, improve, and deliver their goods. Three technology developments that are driving this shift are included in this phenomenon: flexible automation, intelligence, and networking (Korner et al., 2020).

Industry 4.0 is characterized by the way organizations build on the third industrial revolution's work with intelligent, autonomous systems that are powered by data and machine learning. The third and fourth revolutions have been able to coexist because of the emergence of digital solutions and cutting-edge technology, which observers associate with Industry 4.0. These include:

Industrial Internet of Things

- Additive manufacturing (AM)
- Big Data
- Advanced robotics
- Cloud computing
- Augmented and virtual reality (AR/VR)

These technologies help to integrate previously unconnected systems and processes through networked computer systems, which in turn drives the digital transformation of production. Companies may get a wide range of advantages from adopting Industry 4.0, including increased agility, flexibility, customization, and operational performance.

5.2 ADDITIVE MANUFACTURING ROLE IN INDUSTRY 4.0

The key technology behind Industry 4.0 is additive manufacturing (AM). As part of Industry 4.0, 3D printing is evolving into a practical digital technology that offers virtually infinite manufacturing choices, from mass customization to tooling, in practically every industry. AM enables the distributed manufacturing model, which enables components to be created on-demand, by storing design files for them in virtual inventories. By storing digital information rather than actual components, this manufacturing strategy decreases transportation distances, related expenses, and inventory management.

These advantages matter in a variety of industries, including:

Aerospace: AM allows for weight reduction and the creation of intricate geometric parts.

Automotive: Due to its ability to prototype while providing weight and cost savings quickly, additive manufacturing is used to produce a range of materials in the automobile sector.

Medical: This industry is using additive manufacturing parts in an expanding number of applications, particularly for custom-fit implants and devices.

Due to the enabling of customization, lowering of waste, and aiding producers in printing goods where and when they are needed, additive manufacturing is crucial to Industry 4.0. The latter advantage reduces transportation expenses and shortens the time to market.

Making scalable facilities designed for future global commerce is another important contribution additive manufacturing can offer to Industry 4.0. Additive manufacturing and Industry 4.0 work together to produce dynamic, fully customizable, first-to-market solutions.

In addition to being a crucial component of Industry 4.0 now, additive manufacturing will continue to be so in the future. After moving through prototype and small-scale production testing in their factories and establishing additive manufacturing as a core competence, additive manufacturing will be at the center of manufacturing companies' production systems. High levels of investment in additive manufacturing will enable the technology to progress swiftly going forward, particularly with the use of data and artificial intelligence. Additive manufacturing is quickly becoming a crucial element of the factory of the future known as Industry 4.0.

5.3 ADDITIVE MANUFACTURING

Additive manufacturing (AM) is one of the trending advanced manufacturing techniques. Parts are fabricated in layer-by-layer formation (Hölker et al., 2015). Therefore, it has various advantages compared to other conventional processes. The main advantage of the AM technique is to fabricate complex parts with ease (Cyr, Lloyd, and Mohammadi, 2018), while conventional manufacturing processes face many challenges to fabricate complex parts (Hölker and Tekkaya, 2016). The key attraction of the AM is that the part can be manufactured without assembly or with fewer assemblies involved. AM process has given flexibility to build complex parts such as undercut, inside teeth, channels, etc., compared to conventional, there are additive manufacturing processes where they acquire superior properties to the conventional process (Chen et al., 2018; Hölker-Jäger and Tekkaya, 2017). The classification of AM based on material state is shown in Figure 5.1.

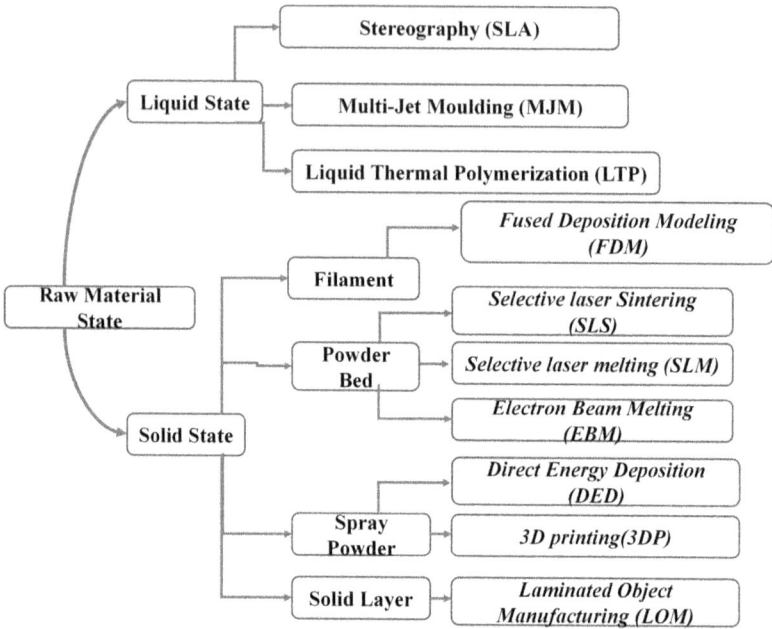

FIGURE 5.1   Classification of AM based on materials state.

The AM process is based on demand by adding material where and when it is needed, and it only fuses at the appropriate locations. Therefore, the process involves minimum wastage of material. The AM technique has been used for various applications and domains, such as medical, aerospace, automobile, marine, etc. (Opatová, Zetková, and Kucerová, 2020; Gokuldoss, Kolla, and Eckert, 2017; Sanjari et al., 2020; Ahmed, 2019). AM also adds its value while creating prototyping, examining the feasibility of the product, new development of the part, optimization of the part as per demand, etc. The advancement of AM techniques has overcome the shortcomings of traditional processes such as functionally graded material, functionally graded properties, making a part of high entropy alloy, etc. (Li et al., 2020; Han et al., 2020).

The numerical modeling capabilities of the AM hold a lot of promise for reducing the time and effort required to produce high-quality items, thus matches the need of Industry 4.0. Industry 4.0 is a strategy that produces higher qualities of parts with complex geometry, however it consumes more time and cost while working with the trial-and-error method to match the need of the industry 4.0 numerical modeling of the AM part used. The AM process takes a longer time and cost to create the product in a ready-to-use state. As a result, it must decrease the effort required to choose and fix the process input to get the intended output. With AM techniques, the reliability of the parts can be managed, thus new materials in AM are released every year (Le Néel, Mognol, and Hascoët, 2018). Now, this can be a great opportunity to revisit the part through the AM material perspective and manufacture most reliably.

5.4 BENEFITS OF ADDITIVE MANUFACTURING IN INDUSTRY 4.0

5.4.1 PRODUCES LESS WASTE AND SCRAPS

When compared to conventional production methods that work by subtracting, additive manufacturing advances in the other manner. In contrast to processes like milling or turning, additive manufacturing involves adding the material required to make objects. It thus generates less garbage and reduces resource waste.

5.4.2 DECREASES PROTOTYPING TIMES AND COSTS

AM has made prototyping products faster, simpler, and less expensive. Other production techniques, like milling, come with high setup and material costs. You can easily develop, test, and make the necessary modifications since prototyping is quicker and less expensive. Additionally, it offers almost immediate evidence of the advancements accomplished.

5.4.3 BUSINESS DIGITALIZATION

For additive manufacturing to be successful, communication between machines, robots, and gadgets must be constant and efficient. Only with the proper digitalization of production processes is this feasible. As a result, businesses increase their investments in digital and IoT, which is necessary for Industry 4.0. It synthesizes the assembly process into a single step.

The streamlining of the manufacturing process, particularly product assembly, is another advantage of additive manufacturing in Industry 4.0. The manufacture of traditional components involves several phases since they are complicated. This increases the price of materials and labor as well as the amount of time needed to make and put together the various pieces. AM, however, enables industry to print the group as a single unit.

5.5 INDUSTRY 4.0 AND AM FOR MAINTENANCE AND SAFETY: A CASE STUDY

This study presents additive manufacturing (AM) as a game-changing technology that can solve metal spare part logistics problems. Figure 5.2 compares the fabrication routes taken by traditional manufacturing versus additive manufacturing. If appropriate authorities approve metals made using additive manufacturing, it is feasible to not only reduce the supply chain for replacement components but also to topologically optimize the final product. Without modifying the setup, it can readily create complex shapes with no financial impact. Even though a few firms have employed AM-built parts in commercial aircraft, such applications are presently only allowed for non-structural parts. There are roughly 30 AM-fabricated components on a Boeing B787 airplane. Nearly 1000 ULTEM 9085 (Fused Deposition Modeling) pieces are used in the Airbus A350XWB. However, the use of AM for structural components requires further research.

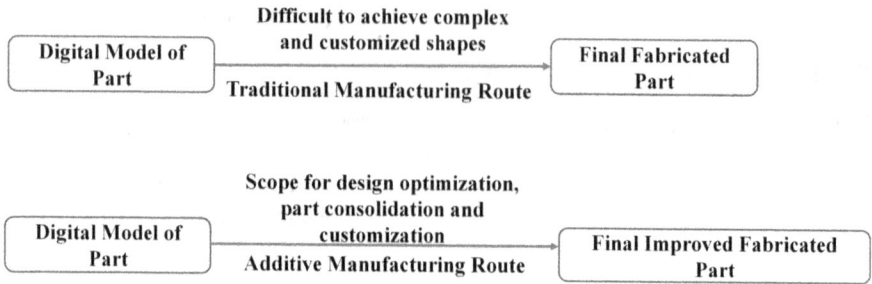

FIGURE 5.2 Comparison between Additive Manufacturing and Traditional Manufacturing.

Studies on maintenance in aerospace under the paradigm of I 4.0, AM and AR are investigated as the processes that can create spare parts and support maintenance, respectively (Ceruti et al., 2019). Aeronautics has been characterized by authors as a difficult and complex discipline, both in terms of design and maintenance. Another difficulty local companies in the civil aviation sector confront is the need to deliver spare parts in a shorter amount of time. I 4.0 and similar technologies have been suggested for use in the field of aeronautics to help with design, maintenance, in-flight structural health, and other processes. While other I 4.0-related technologies can assist in-flight operations, most on-ground maintenance processes can be supported by AM and AR.

MRO (Maintenance, Repair, and Overhaul) operations include procedures like inspection, replacement of broken components, replenishing lubricants and gases (in hydraulic accumulators and damping cylinders), and coating repair. MRO operations are essential for safety in the aviation sector. The national aeronautical agencies such as FAA (Federal Aviation Authority) and EASA (European Aviation Safety Authority) among others, rigorously control these operations as a result. Each Commercial Operator shall establish a Continuous Airworthiness Maintenance Program (CAMP) in line with its Operation Specifications (OpSpecs). The Airworthiness Review Certificate (ARC) is a record confirming the upkeep duties that are completed.

5.6   OBSTACLES TO AM INTEGRATION IN MANUFACTURING SYSTEMS IN INDUSTRY 4.0

Despite the benefits of additive manufacturing, firms are unwilling to use it due to the four key hurdles depicted in Figure 5.3 (Kumar et al. 2020).

The primary barriers to widespread adoption of AM technology are the expenses of integrating the Additive Manufacturing system into the production method and the supplementary costs coupled with materials and post-processing. A Product-Service System (PSS) framework has been developed to conquer this cost obstacle. Within this business structure, industries enter into agreements along with Additive Manufacturing machine vendors for AM installation, associated materials, and maintenance services. Thus, the cost of the equipment is amortized over a longer time. Organizational structures change as a result of the use of AM technology. Additionally, implementing AM demands additional safety precautions and reorganizes the supply

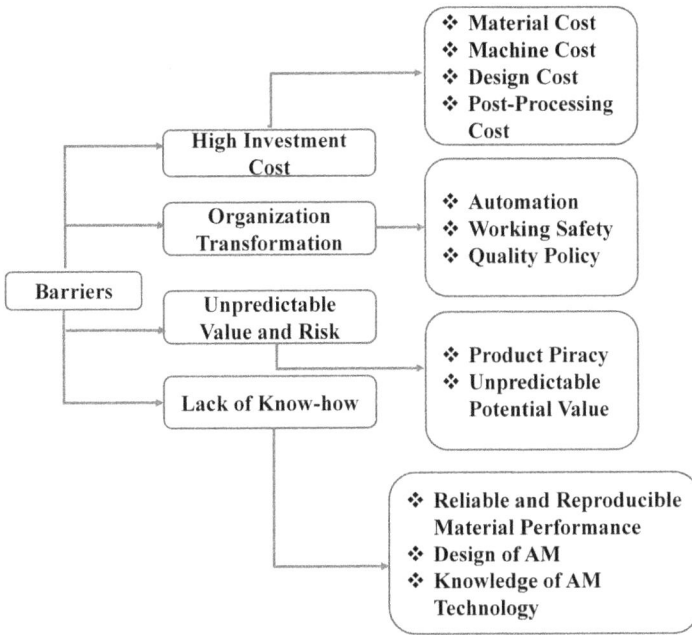

```
                                    ❖ Material Cost
                                    ❖ Machine Cost
                                    ❖ Design Cost
                                    ❖ Post-Processing
                                       Cost
              High Investment
                  Cost

              Organization         ❖ Automation
             Transformation        ❖ Working Safety
                                    ❖ Quality Policy
   Barriers
               Unpredictable
             Value and Risk
                                   ❖ Product Piracy
                                   ❖ Unpredictable
             Lack of Know-how         Potential Value


                                   ❖ Reliable and Reproducible
                                      Material Performance
                                   ❖ Design of AM
                                   ❖ Knowledge of AM
                                      Technology
```

FIGURE 5.3 Obstacles to incorporation of AM technologies.

chain. These modifications result in extra expenses, which deters the system from adopting AM. Some firms would rather continue to utilize the old production system since they are unable to anticipate and understand the advantages of employing the AM method. Manufacturers have highlighted product theft and intellectual property security as additional obstacles to using AM technology. Lack of AM knowledge is another significant obstacle. Companies must increase their technical expertise in two key areas, namely design for additive manufacturing and material performance understanding.

5.7 FUTURE PROSPECTIVE

The faster fish eats the slower fish rather than the bigger fish in today's hugely competitive market environment. Companies strive to produce intricate forms more quickly and inexpensively (Mehrpouya et al., 2019). A product may go from idea to design to development to market in 10% to 50% less time due to AM. Figure 5.4 illustrates the main advantages of implementing AM technology (Kumar et al., 2020).

To provide a quicker and more affordable means of production in an I4.0-enabled smart manufacturing environment, additive manufacturing (AM) technology is still in its infancy and needs more investigation in the areas of suitable Additive Manufacturing materials and design development tools. It is stated that only non-structural parts for aerospace engineering applications use AM build parts in the case study covered in the section above. Structures made with additive manufacturing are still not widely used. One of the causes is the lack of laws governing items made

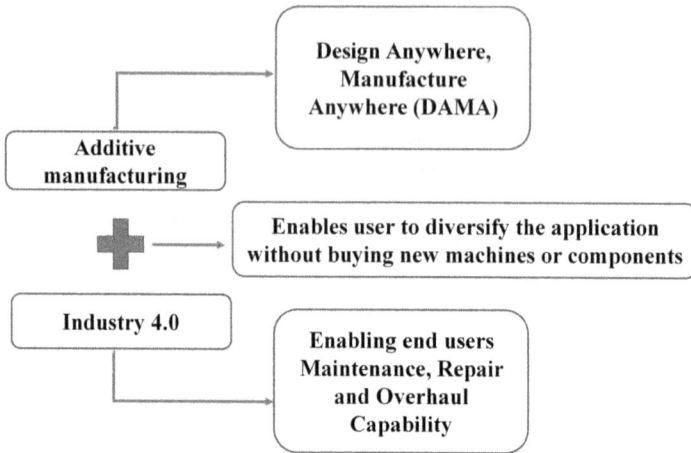

FIGURE 5.4   Combined potential of Additive Manufacturing and Industry 4.0.

using additive manufacturing. Although there has to be a lot of research done before using AM to manufacture structural elements, this tendency will likely alter in the future. Additionally, specialized design tools must be created so that the design flexibility provided by AM may be used successfully. The small number of AM standards that are readily available makes it difficult for businesses to expand their expertise in this area. Two groups have now begun working on this approach. These groups have proposed the publication of a standard series to address concerns pertaining to AM, such as basic ideas, lingo, process classifications, and testing of material. Future AM applications will benefit from these standards, which will act as recommendations.

5.8   CONCLUSIONS

Industry 4.0, a new revolution, is about to begin in the current sector. A decentralized manufacturing process has replaced a centrally regulated production process as the result of the new industrial revolution (I 4.0). In this chapter, I 4.0-related technologies have been briefly explored. The unique attributes of AM, like complexity in shape, material, and function as well as its critical capabilities, such as enhanced design flexibility, component consolidation, and customization for both large and small capacity utilization, have also been discussed in the context of Industry 4.0.

Additive Manufacturing has been acknowledged as one of the key emerging innovations in the I 4.0 concept for the supply chain of spare components. The chapter also contains a case study to illustrate the benefits of AM over conventional production techniques.

This case study proved conclusively that additive manufacturing can easily produce very complex geometries, in the case of lattice-based design, that are usually impossible or very challenging to create using conventional manufacturing processes. This chapter has identified several obstacles that prevent AM from being widely used in modern production systems, despite the fact that it is a "game changer" for the maintenance and parts sector.

The two biggest challenges limiting the integration of additive manufacturing (AM) into industrial processes are the high financing costs and a lack of practical skills. A framework (Product-Service-System) for overcoming the high capital obstacle is also included in this chapter. A research roadmap for materials and model tools for AM within the scope of I 4.0 has been offered in the conclusion.

REFERENCES

Ahmed, Naveed. 2019. "Direct Metal Fabrication in Rapid Prototyping: A Review." *Journal of Manufacturing Processes* 42 (May): 167–91. https://doi.org/10.1016/j.jmapro.2019.05.001

Ceruti, Alessandro, Pier Marzocca, Alfredo Liverani, and Cees Bil. 2019. "Maintenance in Aeronautics in an Industry 4.0 Context: The Role of Augmented Reality and Additive Manufacturing." *Journal of Computational Design and Engineering* 6 (4): 516–26. https://doi.org/10.1016/j.jcde.2019.02.001

Chen, Bo, Yuhua Huang, Tao Gu, Caiwang Tan, and Jicai Feng. 2018. "Investigation on the Process and Microstructure Evolution during Direct Laser Metal Deposition of 18Ni300." *Rapid Prototyping Journal* 24 (6): 964–72. https://doi.org/10.1108/RPJ-01-2018-0022

Cyr, Edward, Alan Lloyd, and Mohsen Mohammadi. 2018. "Tension-Compression Asymmetry of Additively Manufactured Maraging Steel." *Journal of Manufacturing Processes* 35 (August): 289–94. https://doi.org/10.1016/j.jmapro.2018.08.015

Gokuldoss, Prashanth Konda, Sri Kolla, and Jürgen Eckert. 2017. "Additive Manufacturing Processes: Selective Laser Melting, Electron Beam Melting and Binder Jetting-Selection Guidelines." *Materials* 10 (6). https://doi.org/10.3390/ma10060672

Han, Changjun, Qihong Fang, Yusheng Shi, Shu Beng Tor, Chee Kai Chua, and Kun Zhou. 2020. "Recent Advances on High-Entropy Alloys for 3D Printing." *Advanced Materials* 32 (26): 1–41. https://doi.org/10.1002/adma.201903855

Hölker-Jäger, R., and A. E. Tekkaya. 2017. "Additive Manufacture of Tools and Dies for Metal Forming." *Laser Additive Manufacturing: Materials, Design, Technologies, and Applications* 1: 439–64. https://doi.org/10.1016/B978-0-08-100433-3.00017-8

Hölker, Ramona, Matthias Haase, Nooman Ben Khalifa, and A. Erman Tekkaya. 2015. Hot Extrusion Dies with Conformal Cooling Channels Produced by Additive Manufacturing. *Materials Today: Proceedings* 2. https://doi.org/10.1016/j.matpr.2015.10.028

Hölker, Ramona, and A. Erman Tekkaya. 2016. "Advancements in the Manufacturing of Dies for Hot Aluminum Extrusion with Conformal Cooling Channels." *International Journal of Advanced Manufacturing Technology* 83 (5–8): 1209–20. https://doi.org/10.1007/s00170-015-7647-4

Korner, Mario Enrique Hernandez, María Pilar Lambán, José Antonio Albajez, Jorge Santolaria, Lisbeth Del Carmen Ng Corrales, and Jesús Royo. 2020. "Systematic Literature Review: Integration of Additive Manufacturing and Industry 4.0." *Metals* 10 (8): 1–24. https://doi.org/10.3390/met10081061

Kumar, Arun, Gurminder Singh, Ravinder Pal Singh, and Pulak Mohan Pandey. 2020. "Role of Additive Manufacturing in Industry 4.0 for Maintenance Engineering." In *Research Anthology on Cross-Industry Challenges of Industry 4.0*, 235–54. https://doi.org/10.4018/978-1-7998-3904-0.ch013

Li, Yan, Zuying Feng, Liang Hao, Lijing Huang, Chenxing Xin, Yushen Wang, Emiliano Bilotti, et al. 2020. "A Review on Functionally Graded Materials and Structures via Additive Manufacturing: From Multi-Scale Design to Versatile Functional Properties." *Advanced Materials Technologies* 5 (6). https://doi.org/10.1002/admt.201900981

Mehrpouya, Mehrshad, Amir Dehghanghadikolaei, Behzad Fotovvati, Alireza Vosooghnia, Sattar S. Emamian, and Annamaria Gisario. 2019. "The Potential of Additive Manufacturing in the Smart Factory Industrial 4.0: A Review." *Applied Sciences (Switzerland)* 9 (18). https://doi.org/10.3390/app9183865

Néel, Tugdual Amaury Le, Pascal Mognol, and Jean Yves Hascoët. 2018. "A Review on Additive Manufacturing of Sand Molds by Binder Jetting and Selective Laser Sintering." *Rapid Prototyping Journal* 24 (8): 1325–36. https://doi.org/10.1108/RPJ-10-2016-0161

Opatová, K., I. Zetková, and L. Kucerová. 2020. "Particles of Virgin and Re-Used MS1 Maraging Steel." *Materials*, 15.

Sanjari, Mehdi, Amir Hadadzadeh, Ayda Shahriairi, Saeed Tamimi, Hadi Pirgazi, Babak Shalchi Amirkhiz, Leo Kestens, and Mohsen Mohammadi. 2020. *On the Effect of Building Direction on the Microstructure and Grain Morphology of a Selective Laser Melted Maraging Stainless Steel. Minerals, Metals and Materials Series.* Springer International Publishing. https://doi.org/10.1007/978-3-030-36296-6_27

6 Challenges and Prospects of Welding 4.0 Adoption
Implication for Emerging Economics

Raju Prasad Mahto
S V National Institute of Technology, Surat, India

Matruprasad Rout
National Institute of Technology, Tiruchirappalli, India

CONTENTS

DOI: 10.1201/9781003346623-6

6.1   INTRODUCTION

The fourth industrial revolution is defined as Industry 4.0 which is now getting used in various multinational manufacturing companies (Zhikun et al., 2014). Welding is one of the major manufacturing operations where the weld quality is inspected through the various sensors data. The application of Industry 4.0 in welding helps manufacturers to inspect the weld quality through sensor data and a proper feedback system (Debasish Mishra, 2021). The digitization of data and feedback can be applied to any size of weld span and welding type. The concept of Industry 4.0 was first given at the Hanover Fair in 2011 and was later adopted by the German government as one of the strategies for modernizations of manufacturing industries (Xu et al., 2018; Li, 2018). Towards the end of the 18th century, the applications of mechanization in manufacturing changed the process of raw materials extraction from the earth, processing techniques of materials, inspection techniques, etc. This has been defined as the first industrial revolution, as depicted in Figure 6.1. Ultimately it helped industries manufacture products of the best quality in less time, thus helping the economics of countries to grow. The second industrial revolution came into existence at the end of the 19th century when the use of fuels such as gas, oil, and electricity started. It significantly increased the production volume and reduced the lead time. The third industrial revolution came with the use of electronics, computers, and robots in manufacturing processes.

The presence of the internet brought an industrial revolution in manufacturing. In the recent era, the internet helped manufactures to make connectivity from machine-to-machine, machine-to- man, and man-to-enterprise through the web. The connectivity through the web can record the data of all possible segments of manufacturing processes, manpower, robots, etc., in real time. The existence of an abundant amount of data helped industries to make the right decisions at the right time, enabling manufacturers to produce defect-free products of any shape and size in less time.

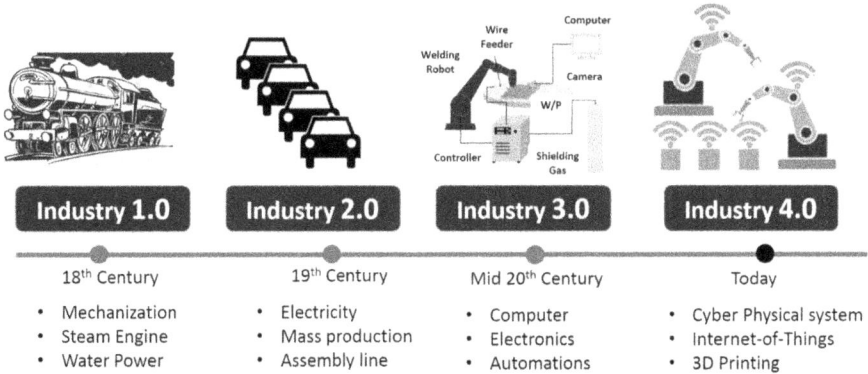

Industry 1.0	Industry 2.0	Industry 3.0	Industry 4.0
18th Century	19th Century	Mid 20th Century	Today
• Mechanization • Steam Engine • Water Power	• Electricity • Mass production • Assembly line	• Computer • Electronics • Automations	• Cyber Physical system • Internet-of-Things • 3D Printing

FIGURE 6.1 Industrial revolutions.

The increase in fuel prices and social awareness across the world have changed the success parameters of many industries (such as automobile, ship-buildings, aeronautic, etc.), to enable use of lightweight materials and deliver strong and multi-functional high performance, etc. To meet such demands, sustainable and ecofriendly manufacturing technologies are the need of the hour. A *joint technique* is one or a combination of the available mechanical, chemical, thermal processes to create a bond between materials with a number of combinations and geometries. In recent years, various welding and joining techniques have been used to tailor the mechanical and metallurgical properties of the products by use of multi-materials. Mechanical joining and chemical and thermal processes are some of the basic processes to establish the bond among the materials. The welding process uses heat, with and without addition of pressure, to produce coalescence of atoms at a temperature below or above the melting temperature of work materials. The welding processes which involve the melting of materials are known as fusion welding techniques, such as electric-arc-based welding techniques, laser-beam welding, electron beam welding, etc. In some welding techniques, no melting of materials takes place. These are known as solid-state welding techniques. Friction welding, pressure welding, cold welding, forge welding, ultrasonic welding, and friction stir welding are some of the well-known solid-state welding techniques. Selection of welding techniques is mainly done when considering the materials being welded, type of production, requirement of negligible dimensional tolerance, chances of formation of defects, etc. However, regardless of the welding process, maintaining the weld quality or producing a weld free from defects is always a challenging task. This makes room for the introduction of automation for welding industries. Automation, in general, can be stated as the use of different mechanisms to operate a system on its own. The main reason for introducing automation is to reduce or eliminate the human interaction with the system so as to increase productivity as well as minimize the error. Welding, in many cases, is being used as the last process in producing a component. Consequently, any undesired event that happens in welding may make the component completely scrap. Implementing automation in welding targets to achieve defect free welds is achieved by monitoring and controlling the welding parameters. For this purpose, robots, along with other contrivances, can be implemented. However, each welding process has its own challenge while implementing the automation. The concept of Industry 4.0 can be applied in the various welding operations, which may help industries make leaner supply chains. However, to implement Industry 4.0 in welding industries, the manufacturers require state-of-the-art facilities related to different tools, sensors, and systems to ensure the defect free products which have been discussed in the following sections.

6.2   WELDING 4.0

The current industrial revolution (i.e., industry 4.0) has given manufacturing industry automation a completely new dimension. It incorporates different digital tools to make the process automated and help in increasing productivity and reducing scraps. Referring to Figure 6.2, industry 4.0 basically includes data acquisition and analysis with the help of cyber-physical systems and the internet of things (IoT) (Li, 2018). The former is related to the controlling of physical objects through computers, whereas the latter refers to the interconnection of various devices and the remote data transfer between them. In short, it can be said that industry 4.0 involves the digitization of all the available knowledge and information. Hence, the manufacturing equipment needs to have different sensors to acquire the data related to the manufacturing process that can be analyzed in real-time. This ultimately helps with faster decision-making and better control of the manufacturing process. Welding 4.0 is in line with industry 4.0 and involves the implementation of sensors, features detection algorithms, a control loop for the detection, control of weld anomalies, etc. In addition, the health of the machine, which affects the weld quality, can be studied by analyzing the acquired signal. The implementation of industry 4.0 will produce smart factories for the welding industries. In order to implement the smart factory successfully, the know-how of the welding techniques needs to be digitalized so that computers can understand and take effective decisions.

In this context, a brief discussion about the difficulties faced during welding by some of the widely used welding processes is given in the following section. Similarly, welding machinery needs to be equipped with automation technologies

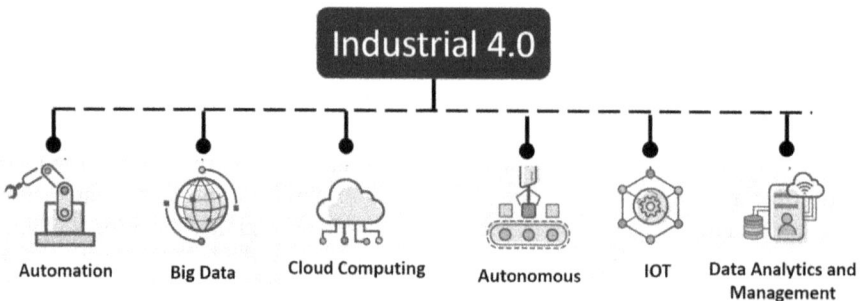

FIGURE 6.2   Fourth industrial revolution.

not only related to data acquisition through sensors but also different communication technologies in order to effectively transfer and analyze the acquired data. Finally, secure infrastructure related to the storage of huge amounts of data is necessary. Automation in the welding process has been in use for a long time. However, in the early days of automation, the focus was on analyzing electrical signals. Towards the end of the twentieth century the focus was shifted to complete automation of the welding technique.

6.2.1 AUTOMATION

Automation in the context of arc welding and manufacturing often refers to the execution of some or all of an operation's steps mechanically or electronically. Partial automation refers to the performance of some tasks manually, whereas full automation refers to the performance of all tasks without operator adjustment (total automation). Many diverse processes can be automated. Equipment may be fixedly automated to handle a single assembly or family of assemblies, or it may be adaptable enough to be easily changed to carry out identical operations on other components and assemblies (flexible automation). There are many advantages to efficiently using automated or mechanical systems in welding. Several advantages include improved productivity, consistent weld quality, predictable welding production rates, decreased variable welding costs, and lower part costs.

6.2.2 BIG DATA

In the welding processes, Big data is typically understood to be a large volume of data (in excess of 1 terabyte) produced at a high rate, with a high degree of variety and veracity per day. It is defined by the four Vs: volume, velocity, variety, and value (Khan et al., 2014). In the welding processes, data can be generated per second or less. The process can be allowed to run for a longer duration of time repeatedly. Therefore, the volume of data in the welding processes can be in the range of petabytes (10^{15} bytes) or exabytes (10^{16} bytes) (Khan et al., 2014). These data could be voltage, current, power, etc., and can generate data at a speed of 10 petabytes per day. Hence, verities of data can be used for defect monitoring.

6.2.3 CLOUD COMPUTING

It has been based on the virtual platform server located virtually, which has been defined as a cloud where data can be stored as well as can be processed. The platform has a CPU, memory, and all other necessary devices, like a physical computer. The stored data in the cloud can be accessed by more than one user simultaneously through the internet. The most widely used cloud systems are Microsoft's Azure and WeldCloud from ESAB, which provide security to the data from any threat. The data about the welding processes and health of the machine components can be transferred to the cloud at a frequency of more than 23 kHz. Later, the data can be used for driving productivity, maintenance, weld quality monitoring, tracing, and documentation. The user can access multiple data, such as arc voltage, arc time-ON,

Arc time-off, number of welds, the weight of the wire feeder, etc., on a dashboard in a much less time.

Cloud computing can act as a remote-control tool or centralized control tool, depending on the applications. In Industry 4.0 enabling welding processes, the data has to be sent to the cloud in the form of alpha-numeric values or images in real-time. Some delays may occur while sending the data, such as high-resolution images in the operations. Therefore, decentralized processing is required, where the storage of data and their processing can be carried out locally. Such computing technologies are known as fog computing. Industry 4.0 requires a reasonably high amount of data from multiple sensors for better predictions of weld qualities. The data of different sensors get fused and then processed for analysis.

6.2.4 AUTONOMOUS

By processing components according to autonomous principles, the entire product line can be automated without requiring that each component be separately coded. The algorithms are based on part geometry, which reduces programming time and makes adding additional parts simple. It requires real-time simulation of the manufacturing process to make process planning automatic, monitoring of the system, and the programming for the robot programming. A sensor system based on vision can be implemented to identify the locations and orientation of different components of the welding equipment. Similarly, sensors attached to the welding torch/nozzle can take care of the errors related to the positioning as well as movement of the workpiece. The sensors can be optimized to adapt the particular welding process and compensate for the errors related to variations in workpiece gap and weld joint volumes. The digitized data available for the joint to be welded, along with the library for welding procedures, can be used to execute the manufacturing simulation. Specially designed and constructed robots like Robot Transport Vehicle (RTV) can make the navigation more accurate, and all the required welding equipment can be carried by the RTV (Mulligan et al., 2005).

6.2.5 INTERNET OF THINGS

The Internet of Things (IoT), or the linking of smart objects to the global internet network, is one of the most intriguing recent technological advances. When used with weldable products, this kind of integration offers astounding possibilities in the field of welding and joining. The opportunities in this industry are essentially endless. IoT in welding, for instance, might significantly raise the caliber of welds, thus raising the caliber and dependability of complex systems in the industrial, aviation, and power generation industries. It has the potential to greatly simplify the reproduction of qualified welding parameters.

6.2.6 DATA ANALYTICS AND MANAGEMENT

It is an important part of Industry 4.0. Essentially Information aids in decision-making improvement, process improvement, error identification, and error prevention (Figure 6.3). These new information technologies will aid in the decision-making and process development of welding processes. There have been many different

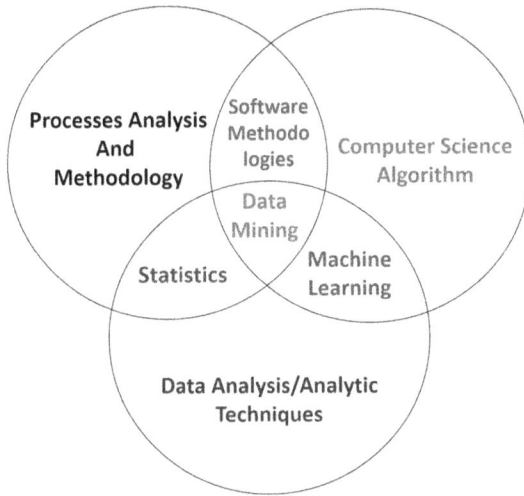

FIGURE 6.3 Important segments of Data Analytics and Management.

techniques applied to welding operations. It is possible to analyze welding processes as a stochastic system with several inputs and outputs. This enables a study from the perspective of data analysis. Processes for data mining, machine learning, deep learning, and reinforcement learning have produced successful outcomes for the analysis and management of different welding processes. The current generation of sensors has improved information capture and information quality, allowing for a huge volume of data that is useful for analysis of these approaches. Descriptive, predictive, and perspective data analysis techniques can be applied with the available data for better decision making.

6.2.7 WELD MONITORING SYSTEM FOR QUALITY MONITORING

The application of industry 4.0 in the field of welding can help the industry develop an automated welding monitoring system. Essentially, it brings digitization to the welding systems on the welding shop floor. In a few industries, the weld quality is being monitored through the recording of the value of voltage, current, and shielding gas flow rate. The system provides an alarm to the welder during the process if the magnitude of current or voltage falls below the required amount. The current system can track the welding procedure but cannot give a guarantee of a defect-free weld sample. A typical weld monitoring system has been schematically given in Figure 6.4.

The weld quality can be diagnosed by using a system that can track the values of multiple process parameters such as voltage, current, gas flow rate, and wire feeding rate in real time. The system must be able to capture the reading multiple times every second. After having data on welding processes over a wider span of time, machine learning/artificial intelligence algorithms can be applied to detect the problems that arise during welding operations. For example, if a manufacturer is only concerned about reducing the internal weld porosity in the welding operations, all the variables have to be identified, which affect the porosity. In a welding operation, values of all

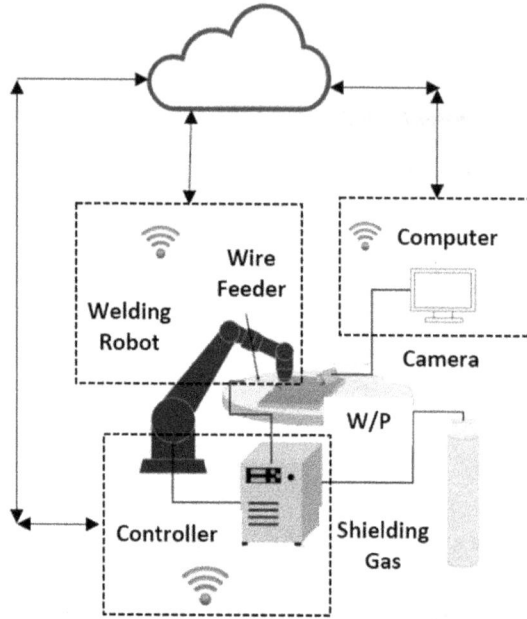

FIGURE 6.4   Inter-connected welding system.

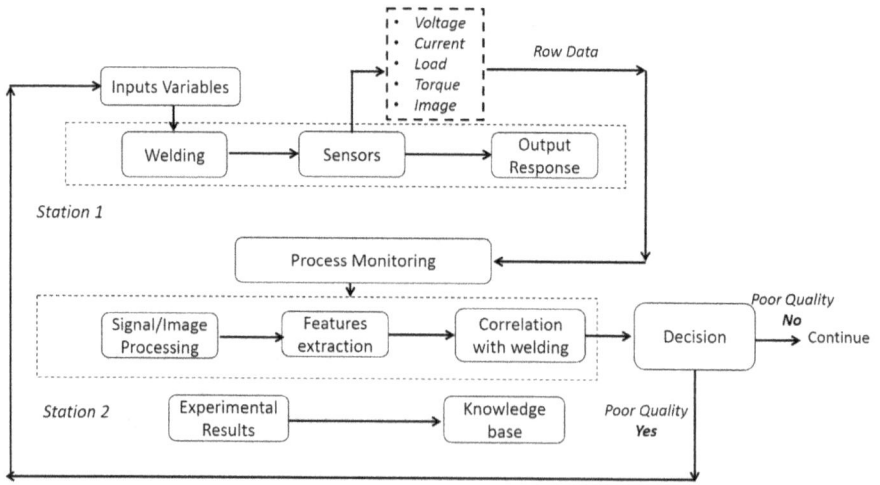

FIGURE 6.5   Closed loop weld monitoring system.

such variables have to be recorded multiple times, for instance, in gas metal arc welding, the magnitude of current, voltage, gas flow rate, and feed rate have significant effects on the gas porosity. Therefore, values of all these variables have to be recorded over a wider span of time in real time. The monitoring system can give warning signals and suggestions to the welder if a suspected defect is detected during the welding operations. The processes have been shown schematically in Figure 6.5.

Industry 4.0 can be applied to any welding technology, such as solid-state welding techniques like friction welding, friction spot welding, ultrasonic welding, resistance welding, gas tungsten arc welding, laser welding, etc. The variables among the different welding techniques can differ among the welding techniques, but ultimately the system will be able to detect the weld quality.

6.2.8 BENEFITS OF WELD MONITORING TECHNIQUES

The application of Industry 4.0 significantly reduces the cost involved with manual weld inspection techniques. Also, rework of manufactures can be reduced as the defects can be eliminated from the welds. The identification of weld defects on components of large size is very difficult through conventional defect monitoring systems. However, the implementation of Industry 4.0 in welding can easily catch defect formation in processes. For example, identification of defective regions in the welding of a wind turbine is very difficult through conventional techniques such as X-ray micro-computed tomography or die-penetration techniques. The application of sensor data can be used to identify the regions or times when the poor weld occurred. The welding inspector can quickly identify the defective areas and repair them at the same time. Therefore, this type of monitoring can save time and effort in avoiding the weld defects. The sensor data can record the welding quality in-process. It can be analyzed for the identification of the root cause of a certain type of defect. This allows an engineer or supervisor to change the process parameters or their values. The identification of root causes of any defect enables engineers to alter the welding procedure, thus allowing enhancement of the weld quality and productivity.

The systems used for weld quality monitoring are able to record and analyze the data, which can be used to develop information report forms. Such information can be uploaded to the manufacturers' enterprise resource planning system. The welding shop or manufacturer can later use such reports to track the cost involved in the welding, such as costs related to electricity consumption, gas, and filler wire. In addition, information available in the ERP can be used for smaller or bigger industries. Therefore, manufacturers are able to enhance productivity by benchmarking best practices in the workshop through analyzing the welding data sheets. The welding information can enable a small-scale welder to work with bigger industries.

6.2.9 IMPORTANT THINGS NEEDED FOR INDUSTRY 4.0

For enabling industry 4.0 welding, a manufacturer needs to invest mainly in the procurement of sensors, power sources, and computers to store, process, and visualize the data. Digital welding systems can also be developed for older welding set-ups, which can record the data of basic variables, such as arc welding, current, arc pulse time on/off, etc. The system can connect all the important segments of welding digitally. Some companies do not have the interest to store and update the in-welding process data for a longer period of time due to reasons like the cost involved in memory, antivirus, etc. Such problems can be solved by the procurement of space in the cloud. Many companies are providing space in the cloud such as Microsoft Azure, etc. The on-site data of welding processes can be stored in the cloud, which can be accessed anytime from anywhere. The cloud providers provide security against any type of virus. The quality of the parts is the top priority of any manufacturing industries regardless of their production capacity. The big manufacturing plant can minimize the work related to re-works due to welding defects by investing in procurements of the sophisticated and advanced welding set-up. Small-scale industry can also maintain the same level of productivity quality by installing sensors to connect all the machines through the web, recording the data, and analyzing them. This is only possible if all the detailed welding data can be accurately recorded. The adoption of industry 4.0 in welding can enable the small-scale industry to make decisions quickly.

One of the important aspects of implementing industry 4.0 in welding is to understand the welding process and the difficulties associated with it. Most welding techniques are based on fusion or solid-state processes, and each of them has some advantages and disadvantages. In order to fulfill the current demand, the use of multimaterials, especially steel, aluminum, magnesium, and thermoplastics, are increasing rapidly in the structural applications of automobile and aviation industries. The introduction of multi-materials in a structure improves the performance and also reduces the total weight. This can be achieved by the proper selection of welding technologies, as it involves materials of different mechanical, thermal, and electrical properties. Since the primary function of automation in welding is to produce defect-free welding by monitoring and controlling the parameters, the difficulties in different commonly used welding processes have been discussed in the following section. This is followed by a brief discussion on monitoring and controlling of welding processes, including the sensors used in welding processes and their importance.

6.3   GENERAL ISSUES AND LIMITATIONS IN WELDING

6.3.1   LIQUID-PHASE WELDING METHOD

Such welding methods apply heat with/without the addition of pressure on a localized area of the materials to be welded. The application of heat melts the material, which leads to fusion and produces the fusion zone that is later allowed to cool to the atmospheric temperature. The fusion may take place with or without the application of external pressure. Resistance spot welding and arc welding are the most commonly used liquid-phase welding techniques in industry.

6.3.1.1   Resistance Spot Welding

Resistance spot welding (RSW) is the most popular welding method in the automobile industries for many decades due to the simplicity, easy automation, speed and low-cost factors of the process. A car body contains more than 5000 spot welds. The nugget in the RSW lies between the two sheets, and the weld is not visible from the outside. In addition, no filler material is required in RSW. In RSW, the electric current is allowed to pass through two electrodes and the workpieces placed in lap configurations. Electric resistance at the contact interface of two sheets helps in producing the heat (Joule's heating affect). Heat is confined to the pointed regions on the workpiece, which is equivalent to the electrodes' point diameter. The heat, along with the pressure applied by the electrodes on the workpiece, helps to form the coalescences of atoms among workpieces, which forms the nugget.

The strength of Resistance Spot-Welded sheets depends on the nugget size, nugget strength, welding current, welding time, hold time, electrode force, etc. In addition, the electrode's diameter and surface conditions of the sheets also affect the nugget strength. Increase in the welding current widens the nugget size and vice versa. Each combination of materials has a certain range of welding current, and if the current is below the threshold value for the given material, then the nugget may not form. On the other hand, at a higher welding current, expulsion of material takes place. The welding time is another important factor, which determines the heating time, hence, the weld nugget size. For the steel body structures, the diameter of an acceptable nugget varies in the range of $4t^{1/2}$ to $5t^{1/2}$, where t represents the thickness of sheet.

In RSW, the welding current-time is represented by the weldability lobes. A wider weldability lobe means a higher range of current will be used in the welding process. This creates problems related to dimensional tolerance. The thickness of sheets and materials determine the weldability lobes. The RSW has been applied in the joining of dissimilar combinations of materials like aluminum-steel, aluminum-copper, and magnesium-steel. However, the welding of such combinations of materials faces a lot of challenges. Most of the difficulties are related to their thermal and metallurgical properties. Some of them are mentioned below.

- Aluminum materials have high thermal conductivity and a significantly lower electrical resistance and melting point as compared to steel. These dissimilarities in the properties require a two to three-time higher electrical current (15–30 kA) and half the welding time for RSW of aluminum, compared to steel (8–10 kA).
- The welding time of RSW for aluminum is about 40–70 milliseconds, which is approximately 1/5th of the welding time requirements for the steel (i.e., around 210 milliseconds).
- Aluminum materials are always covered with its oxide, which is porous in nature. In addition, it has a higher melting point (>2000°C) than the bare aluminum, which creates a lot of difficulties in the joining.
- Surface conditions of sheets are extremely important during RSW of dissimilar materials. Aluminum and magnesium easily react with copper at high temperature. Therefore, in RSW of the mentioned materials with steel leads to faster tool wear, and frequent tool change is required.
- Often lubricants are applied over the metal sheets to improve their formability. The presence of lubricant on the surface of sheets alters the electric resistivity of sheets, which affects the weldability during RSW.

6.3.1.2 Arc Welding Method

In arc welding, heat required for the welding of materials is supplied by the sustainable electric discharge between the two electrodes. The workpiece is acting as one of the electrodes, a wire or rod is another electrode, and both are connected to an electrical supply (Sierra et al., 2008). The wire or rod may be consumable or non-consumable. The consumable electrode serves two purposes: carrying current and supplying filler material after melting to the joint line where the non-consumable electrode only supplies current for electric discharge (Pal et al., 2007). The arc welding takes place at a temperature close to 3500°C, which helps to produce the pool of liquid metal at the joint line. To prevent the liquid metal pool from the chemical reactions with the surrounding gases, inert gases or fluxes are allowed to shield the welding zone.

Arc welding techniques are commonly used in the automobile and shipbuilding industries for the welding of aluminum, mild steel, and coated high strength steel materials. However, the welding of aluminum with steel faces a lot of challenges caused because of the difference in their thermal expansion coefficient, melting temperature, and poor miscibility. The aluminum oxide available on the aluminum sheet also leads to welding defects, as it is porous and tends to absorb moisture from the atmosphere. The moisture consists of hydrogen, whose solubility is less in the solid-state. As a result, when the liquid metal pool solidifies, the hydrogen atoms available on the moisture start to release from the weld pool and cause the porous weld. Sometimes, hydrogen gases get entrapped into the weld at a faster rate of solidification and end up as internal defects in the weld. In addition, presence of the aluminum oxide in the weld reduces the ductility and mechanical strength of the aluminum-steel welds. Therefore, good surface preparations are highly essential in the welding of mentioned dissimilar materials. In the arc welding of dissimilar materials of Al-Steel or Mg-Steel, the metallurgical compatibility of Al, Mg, and Fe atoms are poor, which leads to the formation of brittle intermetallic. The intermetallics are highly brittle, and as a result they easily allow the cracks to propagate under mechanical loading and reduce the strength of the weld structure. Sometimes, the presence of an intermetallic layer at the weld interface leads to the catastrophic failure of the weld.

Often structures are made up of dissimilar material of zinc-coated high strength steel with aluminum and magnesium alloys. The zinc layer reduces the thickness of intermetallics and also improves the corrosion resistance of steel (Schneider et al., 2011). However, arc welding of zinc coated steel with steel at a high welding speed leads to the formation of a porous zinc layer at the weld interface. Welding of dissimilar materials of aluminum alloys requires filler materials as their arc weldability is different. The 5XXX aluminum alloys have better arc weldability than the 6XXX alloys. Therefore, a suitable filler material is required in the welding of different grades of aluminum alloys. Filler materials prevent cracking during the solidification stage. Despite all the precautions, the arc welding of dissimilar aluminum alloys is bound to have a considerable number of thermal distortions, especially for thin sheets because of their high thermal expansion coefficient.

Different grades of high strength steel are also widely used in the structure of shipbuilding and automobiles. Arc welding of Dual phase (DP) steel with martensitic steel produces a wider heat affected zone. As a result, the strength of the weld is significantly lower than the base martensitic steel. The high temperature of arc welding leads to the tempering on the region surrounding the nugget zone in tempered steel, which eventually reduces their static strength. However, fatigue strength does not change appreciably in the welding of DP-Martensitic dissimilar steels.

6.3.1.3 Gas Metal Arc Welding

This welding technique is also known as metal inert gas (MIG) welding, where the arc is confined between the consumable electrode and the workpiece. In this technique, solid wire is fed continuously into the weld pool as a filler material. The filler material composition needs to have a match with the base materials (Zhan et al., 2016). In addition, a mixture of helium and argon gases is supplied into the weld zone to protect the weld pool from the environmental gases. The weld quality in GMA welded samples is also dependent on the process parameters, such as arc current, arc voltage, wire feed speed, electrode travel speed, current density, and preheat temperature. Sometime, faster heat input and cooling in the GMAW produces residual stress and thermal cracks, and hence affects the weld strength. To avoid such problems, preheating of the material to be welded is essential to reduce the residual shrinkage stress of the weld zone.

Welding of dissimilar materials can be accomplished by GMAW, provided the materials have close melting points and good metallurgical compatibility. Poor metallurgical compatibility of materials causes thermal cracking in the heat-affected zone or in the base materials. Sometimes, it also causes formation of microstructure in the weld area, which is highly susceptible to corrosion, and failure at a low weld strength (e.g., GMAW of low-carbon steel with high alloy steel produces a detrimental brittle martensite phase in the weld regions). Similarly, GMAW of aluminum with steel is difficult due to the poor solid-solubility, large difference in melting temperature, and thermal coefficients. The weld zone of an Al-Steel structure is prone to have thick intermetallic compounds, high residual stresses, and weak strength.

6.3.1.4 Gas Tungsten Arc Welding

This welding technique is also known as tungsten inert gas (TIG) welding where the electric arc is established between workpiece and the non-consumable tungsten electrode. Similar to the GMAW, shielding of inert gas is required to protect the weld pool. The process has been used in the welding of dissimilar aluminum alloys, as tungsten electrode has a high melting point (3400°C) and also produces the cathodic cleaning on the work-surfaces. The cathodic cleaning helps to break the aluminum oxide layer, which exposes the arc to the fresh aluminum. This helps in establishing good metallurgical bonding in the weld zone. However, in some cases it requires suitable filler material, which is provided as a separate rod or wire. However, TIG is a slow process and not suitable for dissimilar materials like Al-Steel, Al-Cu, Mg-Steel, etc.

6.3.1.5 Laser Welding

In laser welding, the joint line of the workpieces is exposed to the high energy focused beam. The most commonly used laser sources in the process are CO_2 and Nd:YAG (Neodymium: Yttrium-Aluminum-Garnet) (Cao et al., 2006). The former laser source emits the laser beam at a wavelength of 10.2 μm (near the infrared wavelength), whereas the Nd:YAG produces the laser beam at 1.065 μm. The heat is produced in the joint line by focusing the laser beam over a 0.13–1 mm diameter spot to produce the energy density of 100–110 W/cm^2. Laser welding has remarkable advantages over the conventional welding of dissimilar materials, such as deep, narrow heat-penetration with a narrow heat-affected zone and low heat input. In addition, laser welding can be used to fabricate the structure located at difficult to reach areas by transmitting the laser beam with a series of mirrors.

Despite all, the processes require precise joint fit-up with no gap at the flying surface of the workpiece to be welded. In addition, laser welding requires high initial investment and running costs. The process has successfully welded different grades of steels and titanium alloys. However, welding of dissimilar aluminum alloys is facing a lot of difficulties, as the aluminum has high reflectivity of laser beam. Also, the aluminum alloys have high thermal conductivity, which causes a significant amount of heat loss during the processes. Sometimes, welding of coated steel with aluminum also faces various challenges, such as the coating material can easily vaporize and get entrapped in the weld pool. Later, it can lead to the formation of weld porosity. Laser beams melt the base materials, which vaporize and then dissolve the alloying elements (such as Mg, Si, etc.) in aluminum alloys. As a result this also causes the loss of mechanical strength in the weld area (Mathieu et al., 2007). Added to this, materials like aluminum and magnesium have affinity for hydrogen and oxygen in the liquid state. So, when the laser-welded regions are allowed to cool, internal voids and porosity are observed in the weld. Laser welding of Al-Mg, Al-Steel, and Mg-Steel materials are bound to have internal defects (Cao et al., 2006). In addition, heat in the laser welding is concentrated on a small area. As a result sometimes the weld region has better mechanical strength but has poor elongations at the fracture load. This reduces the formability of laser welded blanks. Therefore, such blanks cannot be used in the designing of complicated shapes. The laser can be coupled with the arc for well-established laser–arc hybrid welding of similar and dissimilar metals (Thomä et al., 2017; Wang et al., 2017).

6.3.2   Solid State Welding Techniques

In solid-state welding, due to plastic deformation a metallurgical bond is obtained between the joining materials. The use of plastic deformation for joining parts potentially offers enhanced accuracy, reliability, and environmental safety. It also creates opportunities to design new products through joining dissimilar materials, which was a major challenge for the fusion or liquid-phase welding processes. The various solid-state welding techniques are forge welding, friction welding, hot pressure welding, cold welding, ultrasonic welding, etc. Among the solid-state welding techniques, friction stir welding (FSW) has gained a lot of popularity. This technique uses a non-consumable rotating tool to join two workpieces without melting the workpiece material (Thomas, 1995). In addition to the plastic deformation caused by the rotating tool, the heat generated by the friction between the rotating tool and the workpiece material helps in the joining the materials. This welding technique has the capability for welding lightweight (Chen et al., 2019) similar as well as dissimilar materials (Mahto et al., 2018; 2020). Else, in the friction welding process, two pieces are moved relatively by means of an upsetting force that heats the two pieces to a plastic-state. The friction welding processes can be rotary friction welding or linear friction welding (Thomas et al., 2002). The ultrasonic welding process uses high frequency ultrasonic vibration to join materials and has a great application in lithium-ion battery manufacturing, carbon fiber reinforced polymer, and aluminum welding, joining of aluminum to copper (Li, 2006) etc.

6.4   MONITORING AND CONTROLLING A WELDING PROCESS

Close monitoring and further controlling the process parameters based on the feedback from the monitoring is crucial to get defect-free welding. Here, the monitoring of the welding process may refer to certain characteristics of that particular process which directly affects the weld on the material, e.g., current and voltage during an arc welding process, tool rotational speed in FSW, etc. Since, welding is a complex process and involves a lot of process parameters, monitoring the welding process is multi-dimensional (Cho, 2008; Jacquin et al., 2004). In addition to the process parameters, there are other factors that need to be considered while monitoring the welding process (e.g., overlapping, mismatch, etc.) Monitoring the welding process can be direct or indirect. Direct monitoring involves certain measurements to characterize the weld, such as hardness, tensile strength, grain size, etc., and is done once the process is over. This can be referred to as an offline approach. Direct monitoring involves dedicated and sophisticated equipment to study the weld quality and needs significant time. The obtained results as qualitative variables are considered for the monitoring of the welding process. Direct monitoring also involves non-destructive methods like ultrasonic testing, radiography, visual sensing, etc.

On the other hand, the indirect monitoring of the process is monitored when it actually happens, i.e., in real-time, and can be referred to as an online approach (Ranjan et al., 2016; Roy et al., 2020). Here, different variables, like current, voltage, torque, force, temperature, sound, vibration, etc., are recorded and analyzed, either online or offline. Different signal processing techniques and machine learning algorithms can be implemented for analyzing the signal and predicting the weld quality. This will further be extended for decision-making, which can be helpful in determining the error when compared to a reference. It will help in building a suitable feedback system to control the process variables during the actual process, i.e., in real-time (Mishra et al., 2018). So, it is very much necessary to know, with respect to a particular welding process, what data/signals need to be acquired, monitored, and controlled.

Sensors, in general, sense the physical quantity that can be further read and analyzed by other devices. So, the physical quantity needs to be acquired as an analog or a digital signal. Depending on the type of welding process, the types of sensors required may vary. Some of the common sensors for welding processes to analyze signals are related to power, current, force, torque, temperature, vibration, acoustic emission, etc. In welding processes like MIG, TIG, RSW, etc., the current drawn from the power source, along with the voltage signals, can be analyzed to relate the weld quality. This voltage-current signal is quite effective to access the weld, as the entire process of welding, including the arc characteristics and transfer of metal, depends on these signals. Sometimes, instead of voltage-current signals, the power consumed for a specific movement, like tool rotation in FSW, can also be useful. However, in this case, the power consumption by the tool movement or spindle motor is a fraction of the power consumed for the entire welding process. Acquiring and processing the electrical data for real-time monitoring is quite helpful, as it does not interfere with the ongoing welding process, and any change in the welding behavior can be figured out from these signals quite comfortably. Similar to voltage-current signals, the force and torque signals can also be helpful in online monitoring of the welding process. However, it is limited to a few welding processes like FSW. Dynamometers can be implemented for this purpose, and the acquired signals can be helpful in identifying defects, tool positioning, etc.

The welding process involves temperature and temperature gradient. The temperature data related to the welding process can be accessed and correlated to the weld development. Different mechanical and metallurgical properties of the weld, which are in general, analyzed post-welding, are temperature dependent. Hence, the temperature during the welding process is quite useful and can be acquired by using different temperature measurement devices used for this purpose, like thermocouples, pyrometers, thermal imagers, etc. However, the exact measurement of the temperature at the weld zone may be a difficult task for some of the welding process.

6.5   CHALLENGES TO THE IMPLEMENTATION OF WELDING 4.0

Automation in welding starts with *digitizing the existing available knowledge*, which is a prerequisite. The easiness of this depends on whether the information can be quantified or not. This is a difficult task as much information available may not be quantitative. The other challenge related to digitization is the automatic accessibility of the data in the future, if required. So, *data storage* is necessary where the data is not only stored for documentation but also for analysis purposes. The data needs to be stored with higher quality as far as possible since the requirement of detailing the data for different algorithms during real-time monitoring may not be the same. These data may be saved in the cloud and accessible to the global automation system of the factory or may be saved locally on any server. However, storage of the data in the cloud requires high-speed internet. Since the concept of industry 4.0 largely depends on data, *security related to data storage* needs to be enhanced in order to avoid the risk of cyber-attack, data corruption, inaccessibility to data, etc. Data security can be enhanced by the installation of high security providing software as well as authentication of welding equipment/hardware. Like storing and security for data, *communication* between devices is equally important, especially when the technology is related to real-time monitoring. Here, communication is not limited to the transfer of data between two or more computer systems. It also refers to other forms of communication like the power supply. All the data should be digitized and communicated to the required microprocessor to perform the required computations. This will further help in generating the required control signals related to the feedback. Establishing an effective communication system is challenging considering the demands of real-time data analysis. So, it is very much necessary to segregate the data/signal required for the communication between the hardware/equipment for effective control of the process and the data required for the documentation.

An effective and defect-free weld cannot be achieved by only implementing industry 4.0, but will also require the proper selection of *process parameters* that define the particular welding process. Since industry 4.0 works with a close network connecting individual systems to make the welding process automatic, the systems should be able to identify the component to be manufactured and also the sequence of operations to be performed. The individual systems need to get or be instructed regarding the welding parameters, and during the process, they need to check the data and modify accordingly, if required. For this purpose, a modern robot designed for the welding process is necessary. These robots need to be equipped with powerful and effective communication and control algorithms as well as microprocessors. In a top to bottom approach, the selection of the proper welding process and corresponding parameters based on the materials to be joined, thickness, joint structure, etc., needs to be identified. This has to be implemented in the very beginning while digitizing the welding knowledge. In addition to the process parameter, the information related to the *positioning of the welding torch/head/nozzle* is an important factor. This can be well defined at the beginning of the welding process; however, as the welding proceeds, the relative position of the torch needs to be along the desired weld path with utmost accuracy. Technology like laser scanners can be implemented, which will further help in getting an adaptive welding process. The application of Industry 4.0 brings automation to the process. However, it has some challenges that include a larger initial outlay of funds than manual welding equipment, the requirement for more precise part positioning and orientation, and the necessity for more advanced arc movement and control systems. As a result, production needs must be sufficient to cover the expenditures of equipment and installation, equipment upkeep, and operator and robot equipment programming training.

Applications of welding 4.0 in the manufacturing sectors can significantly improve product quality and minimize the cost of the processes. However, industries are facing many challenges to implement this concept in developing countries. Some of the major barriers reported by Ghadge and Er Kara (2020) are financial constraints, lack of trained skilled manpower, legal problems, lack of management willpower, insufficient research and development, lack of government policies and support, and poor-quality data. Details of challenges related to the implication of welding 4.0 have been discussed in the subsequent section.

6.5.1 High Investment

Industries that want to execute Industry 4.0 projects would have to promise to increase their anticipated annual capital investments by 50% over the following five years (Raj et al., 2020). This suggests that businesses must not only re-engineer their current strategies but also make a large financial commitment in order to realize the objectives of Industry 4.0. Additionally, Kache and Seuring (2015) pointed out that industry 4.0 enabled manufacturing industries require significant investments in terms of people, processes, and technology. According to Breuing et al. (2016) despite the challengingly high investment requirements, the majority of manufacturing industries are still hesitant to fund R&D connected to Industry 4.0.

6.5.2   Lack of Clarity Regarding Economic Benefit

It has never been clear whether investing in information technology will increase productivity or have positive economic effects. Thus, there is doubt over the accurate assessment of the economic benefits of investing in technology due to the productivity paradox regarding its deployment.

6.5.3   Difficulties in Value-Chain Integration

According to Breunig et al. (2016), overcoming barriers across diverse organizational units is difficult in order to accomplish the seamless coordination required for Industry 4.0. When several firms in the value chain need to be integrated, this type of difficulty is made more difficult. In addition to horizontal value-chain integration, there is a requirement for strong collaboration among value-chain partners in the welding platforms. Majeed and Rupasinghe (2017) claim that the inadequate integration of the IoT in an Industry 4.0 environment is another reason why most businesses fail.

6.5.4   Risk of Security Breaches

Concerns regarding the security hazards of information sharing across channel partners are universally raised by increased connectivity or the intricate links among value-chain partners. The manufacturing industries have the anxiety about losing their data to outside software and service providers, in addition to their concerns about cyber-security (Breunig et al., 2016). As hackers offer significant risks (Lee and Lee, 2020) consider this danger as one of the potential obstacles to IoT implementation.

6.5.5   Poor Maturity Level of Preferred Technology

Technology is changing every day in the world. As a result, the implication of industry 4.0 may lead to several problems in the industries if the standard data security techniques and data processing techniques are not applied. Researchers (Lee and Lee, 2020) bring up the topic of possible chaos when inexperienced, early-stage technologies are applied. Standards, privacy concerns, and data security may not be stable with such technology, and the growing number of untested gadgets may cause turmoil. In addition, the study makes the case that although this wouldn't be a major issue in a disconnected world, it might have a substantial impact on a networked system of technology.

6.5.6   Inequality

The labor market will experience societal unrest as a result of the application of Industry 4.0. The technology will separate the market into low-skill/poor pay and high-skill/high-pay groups, which will cause societal conflict. The consequence of this could be favorable or harmful. Further, it is asserted that Industry 4.0 will increase inequality by widening the difference between those who depend on capital and those who depend on labor, while benefiting intellectual capital owners and their shareholders.

6.5.7 RISK TO EXISTING EMPLOYMENT

Haddud et al. (2017) discussed the possible difficulties related to a reduction in employment if industry 4.0 technology is applied in welding industries. Researchers have pointed out that human displacement is a potential difficulty in a social and organizational setting, even though they do not see it as a challenge for human resources. The developments brought about by Industry 4.0 would likely exacerbate inequality and cause labor market disruption.

6.5.8 LACK OF PROPER REGULATIONS, STANDARDS, AND CERTIFICATION

Literature (Schroder, 2017) reports that because of the non-availability of uniform norms and laws, small and medium-sized businesses (SMEs) are hesitant to adopt Industry 4.0. The lack of standards makes it difficult for the SMEs to focus on joining networks and activities which may create values for the industry. Further, the continuously evolving technology is sometimes a challenge for the regulators and legislators. They may face difficulties in protecting the interests of the customers.

6.5.9 LACK OF INFRASTRUCTURE

The realization of Industry 4.0 requires a comprehensive internet infrastructure. Even in industrialized nations like Germany and France, a key issue is the absence of dependable high-speed broadband for SMEs. Currently, only major players with ample resources have access to this kind of technology (Schroder, 2017). Every channel participant would need to be integrated in order for Industry 4.0 to function, therefore digital infrastructure is a crucial consideration. According to a Penton survey (Hung and President, 2017), 33% of respondents thought that the lack of infrastructure was a barrier to the adoption of IoT. The poll also revealed that many businesses are now working together to construct the essential infrastructure for Industry 4.0, as opposed to competing.

6.5.10 LACK OF DIGITAL SKILLS

According to Breunig et al. (2016), many businesses acknowledge that their lack of the knowledge or abilities needed to fully utilize Industry 4.0 technologies is the main barrier. Authors have suggested that when a company's employees lack the skills to use the software they are purchasing, it paralyzes their capacity to collaborate with the software solution suppliers. Businesses will need to become more data-driven and agile, which calls for a more trained workforce, but underqualified personnel are the second key impediment to implementing Industry 4.0.

6.5.11 CHALLENGES IN ENSURING DATA QUALITY

In the era of digitization, the quality of data is paramount to successfully implementing any digital technology. The data quality, as reported by Zhikun et al., can be assessed by four different components *viz.* correctness, redundancy, completeness, and consistency (2014). All these components are interrelated. A huge amount of data would be generated in the age of fully realized big data, in which the enterprises would need to be connected. The complexity and variety of such data make it challenging to measure the completeness and accuracy of the data. This ups the chance of making erroneous discoveries. Data integrity and consistency are further complicated by frequent data updates and collaboration with numerous parties (Khan et al., 2014). However, if implemented successfully to realize the industry 4.0, industries will be connected at a much higher level than they were in the past.

6.5.12 Lack of Digital Culture and Training

Consistency, completeness, accuracy, and timeliness are the four components of data quality. It becomes vital to have strong internal capabilities inside the team and a culture that supports innovation and is open to experimentation in order to fully realize the potential of Industry 4.0. Additionally, it is asserted that in such a situation, employees with these skills would be in considerably higher demand and much lower supply. Therefore, businesses would need to be prepared to relocate to areas where such workers were accessible. Such tools would foster an internal digital culture and enable the integration of a team made up of subject matter experts from many fields (Breunig et al., 2016).

6.5.13 Resistance to Change

Employee resistance to change in the workplace is one of the obstacles to the adoption of IoT in manufacturing companies. These workers object to utilizing new technologies and the accompanying procedures (Haddud et al., 2017). Researchers such as (Lee and Lee, 2020) assert that various IoT technologies generate a significant amount of personal data, such as home health-related and financial behavior-related data, which many organizations might utilize to leverage their businesses. This is another aspect of resistance. This discourages businesses and people from using IoT because they are concerned about privacy violations.

6.5.14 Ineffective Change Management

Technology has significantly changed the way change occurs during the past ten years. Change used to be seen as transactional and simple to handle, but now it is radical, open-ended, ongoing, and difficult. In other words, it has transformed, making it extremely challenging to execute successfully (Majeed and Rupasinghe, 2017). Since more open-ended systems would be more complicated than ever under Industry 4.0, companies would face a major difficulty in managing change effectively.

6.5.15 Lack of Strategy and Resource Scarcity for Digitization

The technology of Industry 4.0 demands a constant flow of data, both laterally and vertically, within and across businesses. Due to resource limitations, small and medium-scale industries in this situation have greater difficulties than their larger counterparts.

The Innovation Readiness Index (IRI), created by Pierre Audo in Consultants (PAC) in Germany, is discussed by group of researchers (Reinhard, Schrauf, Koch, 2014), who claim that the majority of medium-sized businesses are opposed to using cloud computing. This problem is brought on by senior management because they are more hesitant to implement Industry 4.0 technologies. Additionally, developing a digital strategy to implement Industry 4.0 efforts may be challenging due to misgivings at the top management level regarding the implementation of Industry 4.0.

6.6 SUMMARY

The chapter presents an overview on the challenges that a welding industry faces to implement industry 4.0. Establishing Industry 4.0 for welding can decrease the amount of trial-and-error testing, minimize faults, shorten the time between design and production, and weld products efficiently at a lower cost. The development of a comprehensive, open-source digital twin is made possible by the software and hardware capabilities of the digital age, a global pool of technologically adept and creative workers, and a rich knowledge base of metallurgy, fusion welding, etc. These factors work in concert to make the project both promising and worthwhile from a realistic standpoint. Application of welding 4.0 requires a huge amount of investment. In addition, it requires support from the government policies, along with efforts from the management and workforce of any industry. The manpower has to be trained with various hardware, software, and various analytics techniques. The progress which the manufacturing industries are making in moving forward show the possibility of the complete integration of Industry 4.0 in welding processes.

REFERENCES

Breunig, M., Kelly, R., Mathis, R., 2016. Getting the most out of 2015–2017.

Cao, X., Jahazi, M., Immarigeon, J.P., Wallace, W., 2006. A review of laser welding techniques for magnesium alloys. *J. Mater. Process. Technol.* 171, 188–204. https://doi.org/10.1016/j.jmatprotec.2005.06.068

Chen, K., Liu, X., Ni, J., 2019. A review of friction stir–based processes for joining dissimilar materials. *Int. J. Adv. Manuf. Technol.* 104, 1709–1731. https://doi.org/10.1007/s00170-019-03975-w

Cho, J.-H., 2008. Modeling friction stir welding process of aluminum alloys. *Met. Mater. Int.* 14, 247–258. https://doi.org/10.3365/met.mat.2008.04.247

Debasish Mishra, S.K.P., 2021. Industry 4.0 in Welding. *Mater. Forming, Mach. Tribol. B. Ser. (MFMT)*, Springer Cham 63986.

Ghadge, A, Er Kara, M.M., 2020. The impact of Industry 4. 0 implementation on supply chains. *J. Manuf. Manag.* 31, 669–686. https://doi.org/10.1108/JMTM-10-2019-0368

Haddud, A., Desouza, A., Lee, H., 2017. Examining potential benefits and challenges associated with the Internet of Things integration in supply chains *J. Manuf. Technol. Manag.* 28, 1055–1085. https://doi.org/10.1108/JMTM-05-2017-0094

Hung, M., President, V., 2017. Leading the IoT. Gartner.

Jacquin, D., Guillemot, G., Distribution, W.T., Mehdi, H., Mishra, R.S., 2004. Modelling and Simulation in Materials Science and Engineering An analytical model for the heat generation in friction stir welding An analytical model for the heat generation in friction. https://doi.org/10.1088/0965-0393/12/1/013

Kache, F., Seuring, S., 2015. Challenges and opportunities of digital information at the inter-section of Big Data Analytics and supply chain management. https://doi.org/10.1108/IJOPM-02-2015-0078

Khan, N., Yaqoob, I., Abaker, I., Hashem, T., Inayat, Z., Kamaleldin, W., Ali, M., Alam, M., Shiraz, M., Gani, A., 2014. Big Data: Survey, Technologies, Opportunities, and Challenges 2014.

Lee, I., Lee, K., 2020. The Internet of Things (IoT): Applications, investments, and challenges for enterprises. *Bus. Horiz.* 58, 431–440. https://doi.org/10.1016/j.bushor.2015.03.008

Li, L., 2018. Technological Forecasting & Social Change China' s manufacturing locus in 2025: With a comparison of "Made-in-China". *Technol. Forecast. Soc. Chang.* 135, 66–74. https://doi.org/10.1016/j.techfore.2017.05.028

Li, M., 2006. Interdiffusion of Al – Ni system enhanced by ultrasonic vibration at ambient temperature. *Ultrasonics* 45, 61–65. https://doi.org/10.1016/j.ultras.2006.06.058

Mahto, R.P., Kumar, R., Pal, S.K., 2020. Characterizations of weld defects, intermetallic com-pounds and mechanical properties of friction stir lap welded dissimilar alloys. *Mater. Charact.* 160, 110115. https://doi.org/10.1016/j.matchar.2019.110115

Mahto, R.P., Kumar, R., Pal, S.K., Panda, S.K., 2018. A comprehensive study on force, tem-perature, mechanical properties and micro-structural characterizations in friction stir lap welding of dissimilar materials (AA6061-T6 & AISI304). *J. Manuf. Process.* 31, 624–639. https://doi.org/10.1016/j.jmapro.2017.12.017

Majeed, M.A.A., Rupasinghe, T.D., 2017. Internet of Things (IoT) Embedded Future Supply Chains for Industry 4. 0: An Assessment from an ERP-based Fashion Apparel and Footwear Industry. *Int. J. Supply Chain Manag.* 6, 25–40.

Mathieu, A., Shabadi, R., Deschamps, A., Suery, M., Mattei, S., Grevey, D., Cicala, E., 2007. Dissimilar material joining using laser (aluminum to steel using zinc-based filler wire). *Opt. Laser Technol.* 39, 652–661. https://doi.org/10.1016/j.optlastec.2005.08.014

Mishra, D., Basu, R., Dutta, S., Pal, S.K., Chakravarty, D., 2018. A review on sensor based monitoring and control of friction stir welding process and a roadmap to Industry 4.0. *J. Manuf. Process.* 36, 373–397. https://doi.org/10.1016/j.jmapro.2018.10.016

Mulligan, S., Melton, G., Lylynoja, A., Herman, K., 2005. Autonomous welding of large steel fabrications. *Ind. Rob.* 32, 346–349. https://doi.org/10.1108/01439910510600227

Pal, S., Pal, S.K., Samantaray, A.K., 2007. Artificial neural network modeling of weld joint strength prediction of a pulsed metal inert gas welding process using arc signals 2, 464–474. https://doi.org/10.1016/j.jmatprotec.2007.09.039

Raj, A., Dwivedi, G., Sharma, A., Beatriz, A., Sousa, L. De, 2020. International Journal of Production Economics Barriers to the adoption of industry 4.0 technologies in the man-ufacturing sector: An inter-country comparative perspective. *Int. J. Prod. Econ.* 224, 107546. https://doi.org/10.1016/j.ijpe.2019.107546

Ranjan, R., Khan, A.R., Parikh, C., Jain, R., Mahto, R.P., Pal, S., Pal, S.K., Chakravarty, D., 2016. Classification and identification of surface defects in friction stir welding: An image processing approach. *J. Manuf. Process.* 22, 237–253. https://doi.org/10.1016/j.jmapro.2016.03.009

Reinhard, G., Schrauf, S., Koch, V., 2014. Industry 4.0 – Opportunities and Challenges of the Industrial Internet. www.pwc.de/industry4.0Ind.

Roy, R.B., Mishra, D., Pal, S.K., Chakravarty, T., Panda, S., Chandra, M.G., Pal, A., Misra, P., Chakravarty, D., Misra, S., 2020. Digital twin: current scenario and a case study on a manufacturing process. *Int. J. Adv. Manuf. Technol.* https://doi.org/10.1007/s00170-020-05306-w

Schneider, C., Weinberger, T., Inoue, J., Koseki, T., Enzinger, N., 2011. Characterisation of interface of steel/magnesium FSW. *Sci. Technol. Weld. Join.* 16, 100–107. https://doi.org/10.1179/1362171810Y.0000000012

Schroder, C., 2017. The Challenges of Industry 4.0 for Small and Medium-sized Enterprises.

Sierra, G., Peyre, P., Deschaux Beaume, F., Stuart, D., Fras, G., 2008. Galvanised steel to aluminium joining by laser and GTAW processes. *Mater. Charact.* 59, 1705–1715. https://doi.org/10.1016/j.matchar.2008.03.016

Thomä, M., Wagner, G., Straß, B., Wolter, B., Benfer, S., Fürbeth, W., 2017. Ultrasound enhanced friction stir welding of aluminum and steel: Process and properties of EN AW 6061/DC04-Joints. *J. Mater. Sci. Technol.* 34, 163–172. https://doi.org/10.1016/j.jmst.2017.10.022

Thomas, W.M., 1995. Patent Friction Welding Thomas TWI. United State Pat.

Thomas, W.M., Nicholas, E.D., Watts, E.R., Staines, D.G., 2002. Friction based welding technology for aluminium. *Mater. Sci. Forum* 396–402, 1543–1548. https://doi.org/10.4028/www.scientific.net/MSF.396-402.1543

Wang, H., Feng, B., Song, G., Liu, L., 2017. Laser–arc hybrid welding of high-strength steel and aluminum alloy joints with brass filler. *Mater. Manuf. Process.*, 1–8. https://doi.org/10.1080/10426914.2017.1364762

da Xu, L., Xu, E.L., Li, L., 2018. Industry 4.0: state of the art and future trends 7543. https://doi.org/10.1080/00207543.2018.1444806

Zhan, X., Li, Y., Ou, W., Yu, F., Chen, J., Wei, Y., 2016. Optics & Laser Technology Comparison between hybrid laser-MIG welding and MIG welding for the invar36 alloy. *Opt. Laser Technol.* 85, 75–84. https://doi.org/10.1016/j.optlastec.2016.06.001

Zhikun, C., Shuqiang, Y., Shuang, T., Hui, Z., Li, H., Ge, Z., Huiyu, Y., 2014. The data allocation strategy based on load in NoSQL *Database* 517, 1464–1469. https://doi.org/10.4028/www.scientific.net/AMM.513-517.1464

7 An Approach to Friction Stir Additive Manufacturing of Light Weight Metal Alloys

Prakash Kumar Sahu

National Institute of Technology Agartala, Tripura, India

Anmol Bhengra

KIIT Deemed University, Bhubaneswar, India

Jayashree Das

Indian Institute of Technology, Guwhati, India

CONTENTS

7.1 INTRODUCTION

The Industrial Revolution can be divided into four different stages. The first industrial revolution saw the utilization of steam power to increase productivity, which

DOI: 10.1201/9781003346623-7

was minimal previously due to the complete dependence of manufacturing on human labor. The second industrial revolution saw the use of electricity for mass production and a decrease in average manufacturing time. The third industrial revolution introduced the concept of automation in the manufacturing process by making use of computer technology (Yu et al., 2015). The fourth industrial revolution, which we are in currently, has taken the idea of automation to a whole new level (Horvath et al., 2019). The review research in this stage of the industrial revolution explores the full potential of machines to make them interact so that there is coordination among different machine parts used to manufacture certain components. These include the advent of Artificial Intelligence (AI), making the computer smarter and helping it learn complicated jobs, recognize complex patterns with faster computer processing, Robotics that makes the manufacturing process fully automated, and one of the most important inventions of the fourth industrial revolution, Additive Manufacturing Technology (AM). Also, the FSAM is one of the advanced AM processes in the present scenario.

Additive Manufacturing is the industrial name for 3D printing, which consists of a number of manufacturing technologies that create 3-dimensional objects with a layer-upon-layer manner of material deposition (Ahluwalia and Mahto, 2018). This is an improved manufacturing technology compared to the conventional manufacturing technology that involved subtraction of materials (Prakash et al., 2018) in the form of chips. In traditional manufacturing, the parts need to be manufactured separately and assembled, which takes more inventory space and time to produce a finished product. Whereas, in additive manufacturing, the parts can be manufactured as a single unit within a shorter period of time, leading to improved structural integrity and productivity. The process of additive manufacturing involves 6 stages, namely: (i) Design a 3D CAD model of the component, (ii) Tessellate a 3D model for input, (iii) Slice the 3D CAD model into layers, (iv) Determine the positions at which the material needs to be deposited, (v) Create the physical model, (vi) Post-processing of the finished model to meet the required dimensional tolerance.

Additive manufacturing is classified into seven types by the ASTM, namely, VAT Polymerization, Material Extrusion, Material Jetting, Binder Jetting, Powder Bed Fusion (PBF), Sheet Lamination, and Directed Energy Deposition (DED) (Chua et al., 2017). AM can also be classified on the basis of either fusion based AM technology or Solid-State AM. Under fusion state AM there are components that are build layer-by-layer by locally melting a metallic powder, using an electron beam as the source of heat. Processes like Power Bed Fusion and Directed Energy Deposition are some examples of fusion-based AM. Then, there are the Solid-State AM techniques such as Friction Stir Additive Manufacturing (FSAM) and Cold Spray Additive Manufacturing that do not involve the melting of material at any stage of their operation (Derazkola et al., 2020). The classifications of additive manufacturing is represented in the Table 7.1.

The different types of the Additive Manufacturing Process have their own unique characteristics with advantages depending upon the kind of method they go through. Binder Jetting uses a binding or binder for joining together the powder metals which have been spread over by a roller. This gives the binder jet process speed of manufacture. It can also involve a variety of materials, though the structural strength of the parts produced by this process are low and thus cannot be used for demanding

TABLE 7.1

Classification of Additive Manufacturing Processes

Classification of AM by ASTM	Classification Based on State of Material	
	Fusion Based	**Solid State**
VAT Polymerization	Powder Bed Fusion	Friction Stir Additive Manufacturing
Material Extrusion	Directed Energy Deposition	Cold Spray Additive Manufacturing
Material Jetting		
Binder Jetting		
Powder Bed Fusion		
Directed Energy Deposition		
Sheet Lamination		

components. The powder bed fusion process uses a laser light source or electron beam to melt the plastic or metal powder, which then solidifies after cooling in the desired pattern. The process also has a low cost of machining. A variety of materials can be used, but again this process is unable to produce good structural components and takes a lot of time and energy. Directed Energy Deposition (DED) provides a better solution for making structural components, as it involves the use of a focused thermal energy source like a laser to melt and deposit powder or wire-based material (Wolff et al., 2017). This process is able to produce complex shapes and reduce the wastage of materials. However, the parts produced by this process have poor surface finish, and the overall cost of setting up the system is high. Sheet lamination uses stacking and laminating of thin sheets of material bonding and is finally shaped with the help of CNC machining. This technology can produce parts with the least number of details or complexity, but the good thing about it is that it is budget friendly and has faster manufacturing time. The three types of sheet lamination are Selective Deposition Lamination (SDL), Laminated Object Manufacturing (LOM), and Ultrasonic Additive Manufacturing (UAM) (Honarvar et al., 2020). Some disadvantages of this technique include that it gives a limited material option, hollow parts are difficult to manufacture, it leaves waste material, and the bonding of the structure produced depends upon the method of lamination used. In the material extrusion technique of additive manufacturing, semi-solid material is pushed through a nozzle, bonding to the previously deposited layers by solidifying to build a part (Lee et al., 2017). The material is melted and liquefied in a heat chamber in the form of a filament. In this technique, the parts that are produced can have a wide selection of print materials that have low initial and running cost, being a low temperature process. In case of material extrusion, 3D parts are constructed by making use of thermoplastic or composite material. This method uses a filament of plastic which is fed through a nozzle as extrusion and deposited on a platform in a layer upon layer manner. This process can fabricate a variety of materials, and has low initial and production cost. Still, this technology has many disadvantages, such as weakness in part strength along the z axis of the component. The components produced also have issues related to warping, and other effects that are caused because of rapid temperature changes. The materials used for this process can also be toxic and unhealthy (Figures 7.1 and 7.2).

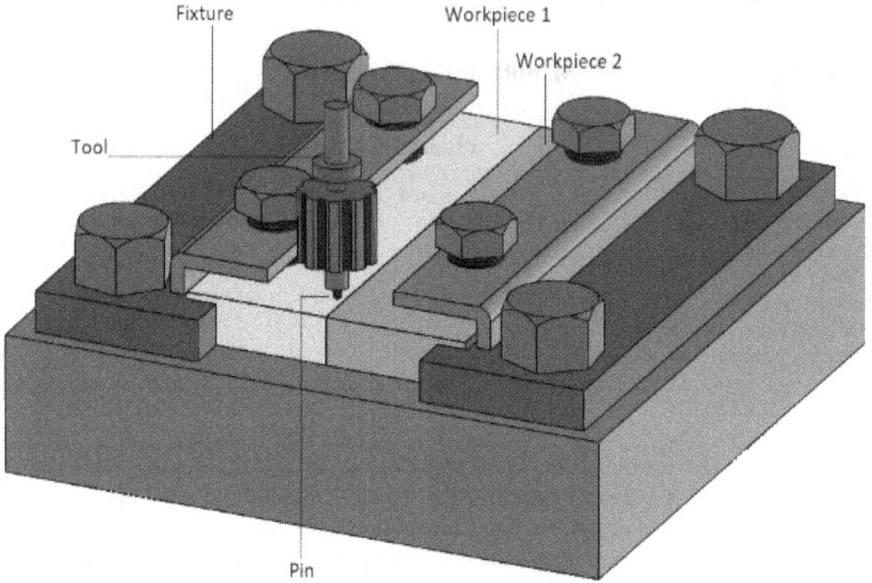

FIGURE 7.1   Friction stir welding and processing setup.

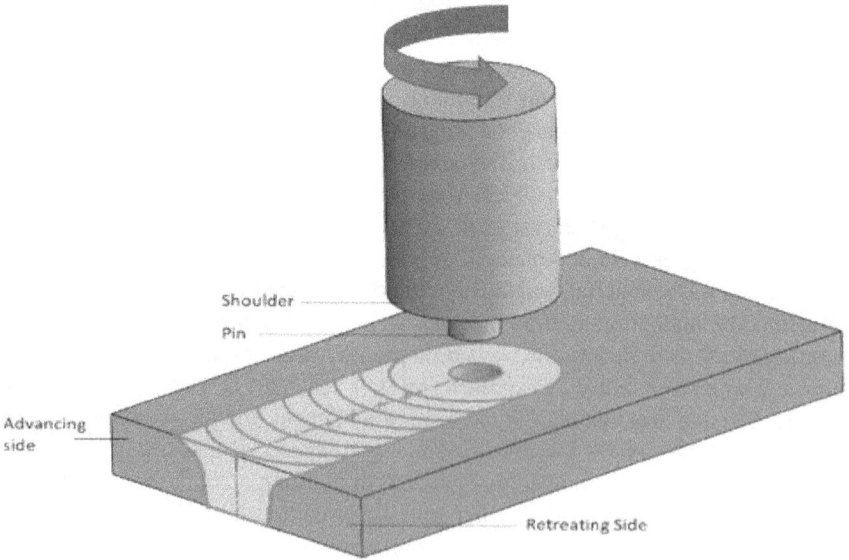

FIGURE 7.2   Schematic of the friction stir processing.

Material Jetting is another type of additive manufacturing process that makes use of wax-like material for deposition on the build platform in a desired pattern. Afterwards, the deposited material is cooled and solidified, making the component layer by layer. This AM process is good for achieving outstanding accuracy and surface finish, also the parts produced are good to use in casting patterns. Although this technology also has some disadvantages, such as the parts are weak and fragile because of the use of the wax type (semisolid) material and the building process is considerably slow. The last is the VAT Polymerization technology, which involves the solidification of a liquid photo sensitive resin, but selectively exposing it to radiation, which is usually ultraviolet light. The photosensitive resin used in epoxy is based to ensure low shrinkage, with added acolyte to speed up curing time. The three most common types of this technology are digital light processing, stereo lithography, and continuous digital light processing (Ali et al., 2020). This technology provides more dimensional accuracy and the highest finish, is quick, and has a large build area. Disadvantages of this technology include being expensive, requiring post curing so that the parts can be used for structural use, and also that it is limited to photo resin materials.

Additive manufacturing technology can also be classified into fusion based AM and solid-state AM technologies (Derazkola et al., 2020). Fusion based AM comprises Powder Bed Fusion and sintering by a laser or an electron beam energy source, or the melting of powder or wire as welding to build up parts. Some types of PBF are: Selective Laser Sintering (SLS), Selective Laser Melting (SLM), Direct Metal Laser Sintering (DMLS), and Electron Beam Melting (EBM) (Ali et al., 2020). These methods are appropriate for different types of application such as medical implants or structural aerospace parts. In the solid-state process there is no melting of the parts involved, however products are fabricated by friction, pressure and velocity. The best example of solid-state AM processes are Friction Stir Additive Manufacturing process (FSAM), Cold Spray Additive Manufacturing process and Sheet Lamination (Jahangir et al., 2018). In cold spray additive manufacturing, highly accelerated powder particles are deposited on the substrate with the help of a diverging nozzle. This process reduces the size of the HAZ (Heat Affected Zone), thus reducing the stresses being induced on the workpiece. The concept of Friction Stir Additive Manufacturing and Friction Stir process came after the invention of the Friction Stir Welding process, which was invented by Wayne Thomas at TWI ltd. in 1991.

7.2 ORIGIN OF FRICTION STIR ADDITIVE MANUFACTURING (FSAM)

The FSAM technology was first introduced around 2006 when Boeing and Airbus demonstrated how structures can be fabricated at a quicker speed with minimal waste material (Palanivel et al., 2015a). In FSAM a rotating tool is fitted with a custom designed shoulder and pin that is meant to be injected into the surface of sheets to be joined along traverse line. Similarly, many layers of workpieces are joined layer by layer to construct a component of a desired shape and size, as shown in Figure 7.3.

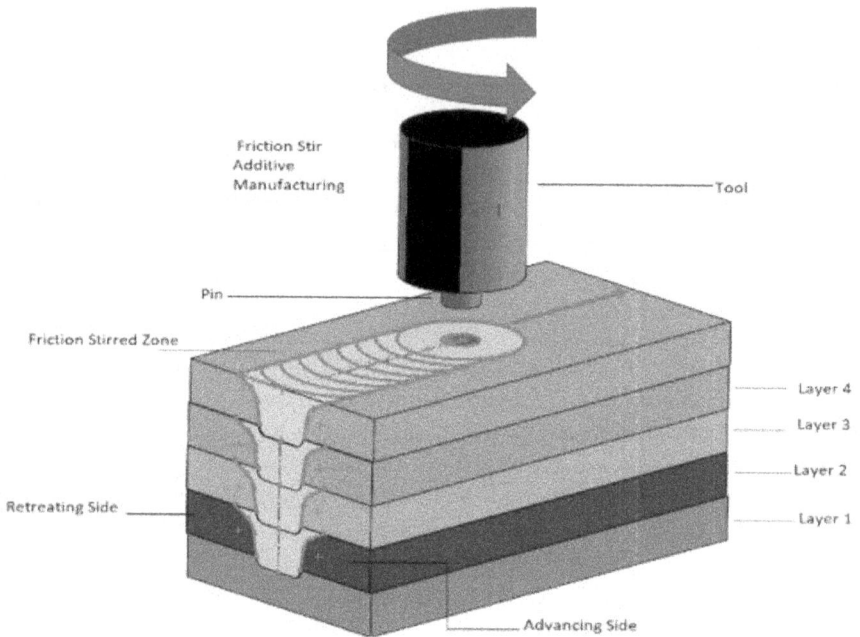

FIGURE 7.3 The schematic of Friction Stir Additive Manufacturing.

The joints that are produced by this process are in solid state and do not require melting at any stage, thus eliminating the emergence of defects that are usually caused by melting, such as cracking, porosity, anisotropy, and residual stresses due to the temperature of the process. Also, the FSAM process is able to produce parts that have finer grain microstructure as opposed to the other fusion based techniques.

7.2.1 Parameters Affecting FSAM

- **Tool Rotational Speed**: It is the speed at which the tool rotates on its own axis when it is inserted inside the workpiece and is measured in rev/min (RPM) (Pareek et al., 2007). It has generally been seen that when a FSAM process involves a tool rotating at high speeds the component exhibits the formation of discontinuities, whereas less tool speed results in better flow of plasticized material across from the retreating side to the advancing side of the workpiece.
- **Traverse Velocity**: It is the speed at which the tool goes along the predetermined line of travel to perform the process of FSAM. It can be measured in mm/min.
- **Tool Plunge Depth**: It is the depth to which the shoulder of the tool is inserted inside the work piece such that the shoulder of the tool should be in firm connection with the surface of the workpiece (Jain et al., 2018).
- **Tilting Angle**: It is the angle at which the tool is tilted with respect to the horizontal surface of the body of the workpiece.

- **Added Material**: The added material to the friction stir process may be in the form of powder, foil, flakes, fiber, etc., which can be compatible with the stirring process. Some external material is added to the base plate of the friction stir process, thus this is called friction stir additive manufacturing process.

7.2.2 Types of FSAM Process

There are four main ways in which the FSAM process can be carried out for manufacturing of components. These are:

- **Powder or feed rod technique**: As shown in Figure 7.4. A hollow cylindrical tool is used in this process that contains a powder or a consumable rod which is fed to the surface of the substrate. Thus, the rod gets deposited on the substrate (Stelt et al., 2013).
- **Cladding or Sheet Lamination technique**: In this technique, sheet-like structures are deposited onto the surface of a component for various purposes such as covering the component with a protective coating.
- **Functionally Graded Materials**: In this method of FSAM technique, a groove is made into a workpiece, then another metal is added by stirring it inside the first one. Then this process is repeated until a workpiece with the desired variation in mechanical properties is achieved.

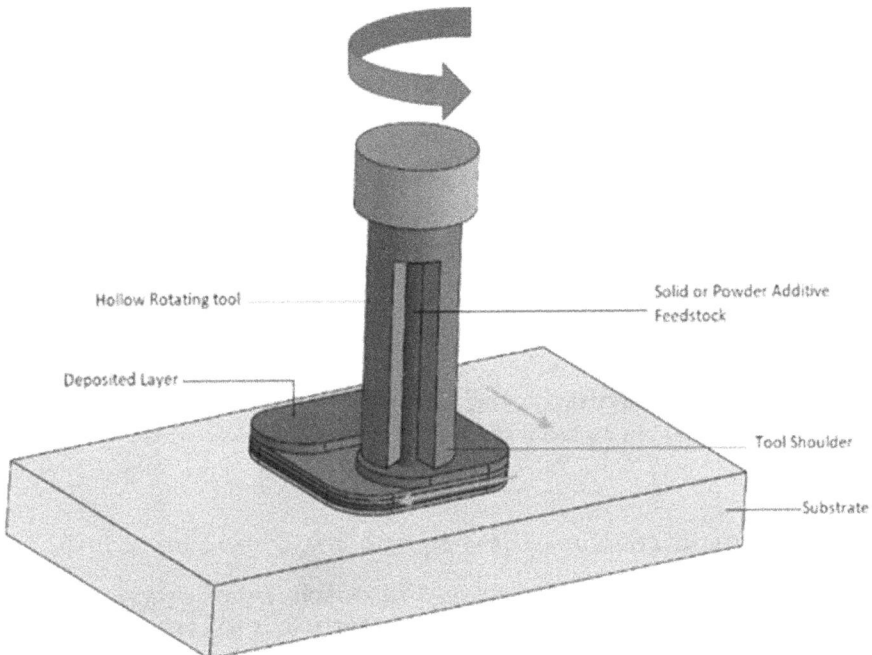

Hollow Rotating tool

Solid or Powder Additive Feedstock

Deposited Layer

Tool Shoulder

Substrate

FIGURE 7.4   Schematic of additive friction stir deposition using a feed rod technique.

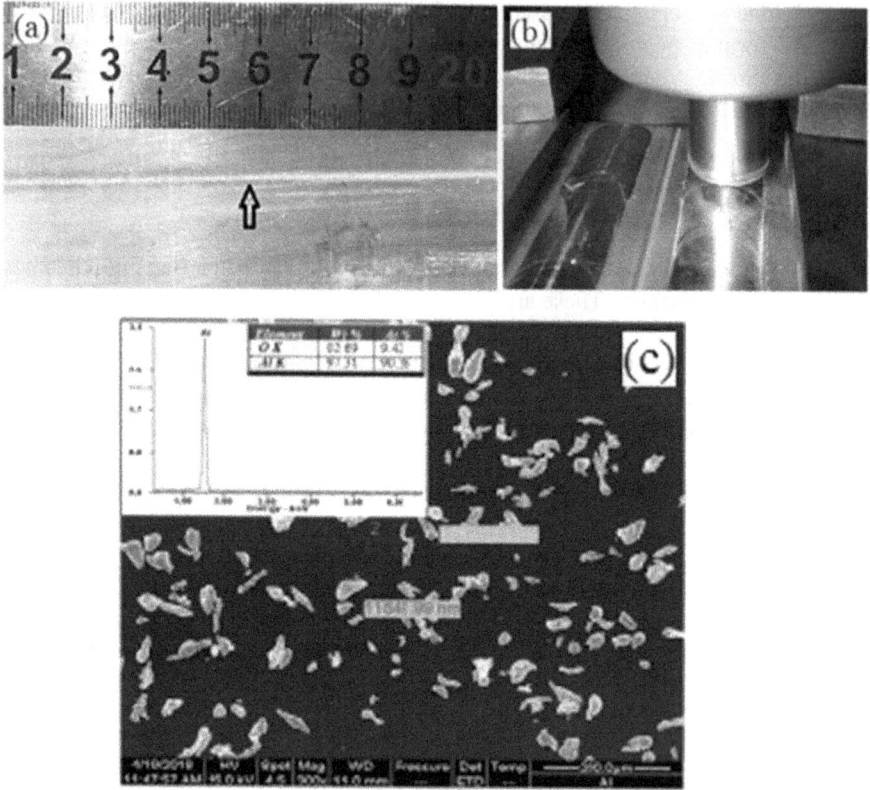

FIGURE 7.5   (a) Illustration of the alloying powder packed inside the groove (b) Single pass for closing the groove using pin less tool, (c) Used Al powder for the FSAM.

- **Friction Surfacing Technique**: This is another sort of deposition technique where a consumable rod is fitted at the end of a rotating tool and is made to frictionally rotate on the surface of substrate where the material is to be deposited in the desired pattern.
- **Coating**: There is also a technique of friction stir modification where coatings which have been deposited by other methods such as cold spray can be altered by FSAM.
- **Friction Stir Selective Alloying**: This also can be considered as new age process to modify the metal surface, as shown in the Figure 7.5.

7.2.3 Different Zones in a FSAM Process

With the passage of a tool through the material workpiece, the material experiences different temperatures at different locations at a given time. These locations can be classified in the form of different zones to better clarify the thermal effect of the tool on the workpiece as the friction stir additive manufacturing process is being carried out. The different zones of the FSAM process are (i) **The Preheat Zone**: The preheat

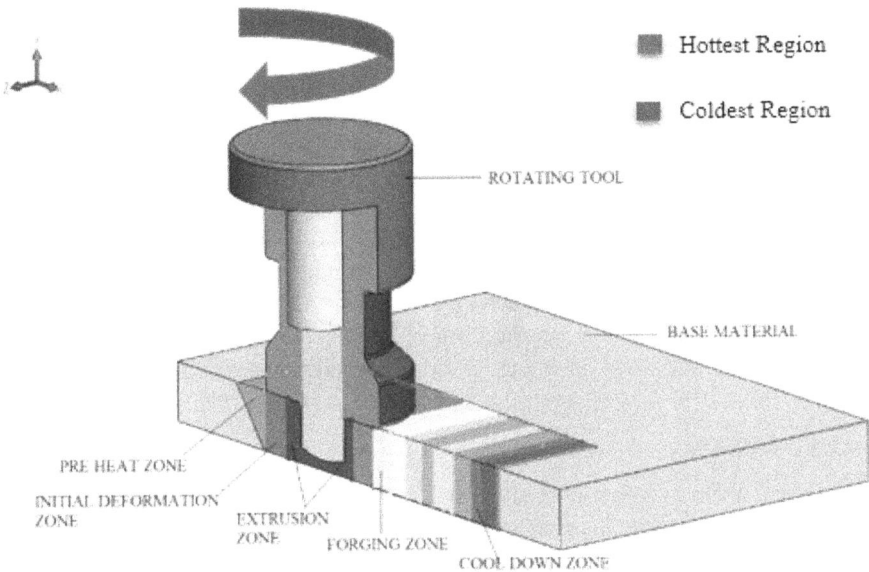

FIGURE 7.6 Schematic of the different zones in case of additive friction stir deposition.

zone lies ahead of the tool and is about to be deformed as the tool will reach it. This zone experiences some amount of heat as the tool is close to position. (ii) **Initial Deformation Zone**: In the initial deformation zone, the mixing of the material takes place by the action of the tool shoulder following the flow path of the material. This zone also experiences some amount of heat as it is about to be plasticized by the tool (Puleo et al., 2016). (iii) **Extrusion Zone**: The zone that surrounds the pin is known as the extrusion zone. The material in the zone gets pushed around by the pin. The pin pushes the material from front to back and simultaneously pushes it in the downward direction towards the bottom of the weld (Li et al., 2013). (iv) **Forging Zone**: The zone after the extrusion zone is known as the forging zone. The material that comes out as the pin is moving is forced into the gap present in between the hardened material and the pin and is also compressed by the application of the load provided by the shoulder (Puleo et al., 2016). (v) **The Cool Down Zone**: This is the final zone. where the plasticized material starts to cool down and starts to become the weld that will become the joint in joining two or multiple layers of materials. The schematic of the different zones in case of the FSAM are represented in the Figure 7.6.

7.3   MATERIAL COMPATIBILITY WITH FSAM PROCESS

To fully utilize the potential of the FSAM manufacturing process, we first need to establish the metals or materials that can be used for this process because certain metals have a very high melting point, making them not ideal to use in FSAM since they can damage the tool by causing stresses that the tool won't be able to handle. Also, there has been a continuous demand to reduce the weight of components for use in the aerospace, marine, and automobile sectors so that they become more energy efficient. Thus, for this purpose, certain metals and materials have been ideally chosen for the purpose of the friction stir additive manufacturing process. This includes alloys of nonferrous metals such as magnesium, aluminum, copper and its alloy, zinc, Inconel, zirconium, Inconel and titanium, etc. These metals have been found to exhibit characteristics that are able to fulfill the requirements of being lighter in weight, having higher strength to weight ratio, and higher corrosion resistance capabilities, making them an ideal substitute for steel that is used in different engineering applications. Recently, it has been observed that the use of lighter materials enhances the part performance in various engineering purposes.

However, for the wide application of light weight metals like magnesium and aluminum, different research focused on the Mg and Al alloy. In this review, we will discuss in detail the use of magnesium and aluminum alloys and how these will have compatibility with FSAM, owing to their different advantages as discussed below.

7.3.1 MAGNESIUM ALLOYS

The development of magnesium alloys has seen an increase in recent years owing to their increase in demand in the aerospace industry for their being a lightweight material and other characteristics as follows.

- Mg alloys are light in weight, making them suitable for various engineering application.
- Mg alloys have low density.
- Mg alloys exhibit good high-temperature mechanical properties.
- They have good to excellent corrosion resistance.

7.3.2 ALUMINUM ALLOYS

Al alloys have mechanical properties similar to magnesium alloys and can be used for various engineering applications, owing to their following characteristics.

- Al alloys exhibit good strength, and some alloys can even have strength up to 300 MPa and are stronger than some steels.
- Al alloys have high strength to weight ratio. Al has the ability to recover under the application of static and dynamic loading conditions.
- Al has strength at low temperatures.
- Similar to magnesium alloys, aluminum alloys can have good corrosion resistance
- Al is non-toxic in nature and also has good electrical conductivity.
- Al is nonmagnetic and non-combustible.
- Aluminum's non sparking characteristics make it an ideal material for products used with highly flammable or explosive substances and atmospheres.

7.4 MICROSTRUCTURAL ANALYSIS OF MAGNESIUM ALLOYS PRODUCED BY FSAM

The present review work aims to determine the microstructural change of the Mg alloy after the FSAM process. For this reason, let us consider a Mg alloy like WE43 and AZ31B alloy. The chemical composition of WE43 is: Yttrium 3.7%, Neodymium 2.2%, Rhenium 0.96%, Zinc 0.51%, with the balance being Mg alloy. Similarly, AZ31B composition is: Al 3.0%, Mn 2.0%, Zn 1.0%, Cu 0.05%, Ni 0.005%, Si 0.1%, Fe 0.005% and Mg. Microstructural analysis is necessary to observe the changes in the grain size and its effect on the other properties after the FSAM.

A microstructural analysis is accomplished in order to determine the changes in the microstructure in a metal alloy. This observation is made at different levels of magnification, depending on the nature of observations to be made. It helps to evaluate the various states of supply in a material and presence of metallurgical defects that could have emerged in the considered material. This analysis is generally done from a section of the piece to include the observation at the core and on the surface in regions that are thick and thin areas so as to find out the variation in characteristics of the material with the changes in characteristics of the material with the change in thickness. This analysis can be done on samples that are small in dimension, although it would be more effective if it was done on the most representative areas.

There are mainly three process parameters that affect the size of grain structures:

- **Rotational Speed**: As discussed later, it is established that rotational speed can greatly affect the formation of fine microstructures in an alloy due to the pulsating effect of the rotating pin and shoulder contact.
- **Traverse Speed**: Similarly, traverse speed can also alter the mechanical properties and microstructure formation in a material (Shunmugam and Kanthababu, 2018). The traverse speed affects the amount of heat generation, and that directly affects the grain recrystallization and enhancement of the mechanical properties.
- **Tool Geometry**: It has been observed that different types of tool design have varying effects on the stirring effects, which in turn lead to the variation in fluidity and grain size of the plasticized material (Gao et al., 2021). The different types of the pin geometry are represented in the Figure 7.7.

In this part of the review paper, we will get know about the various microstructural changes that take place before and after the FSAM process is carried out on a number of magnesium alloys. The main aim of this study is to establish the fact

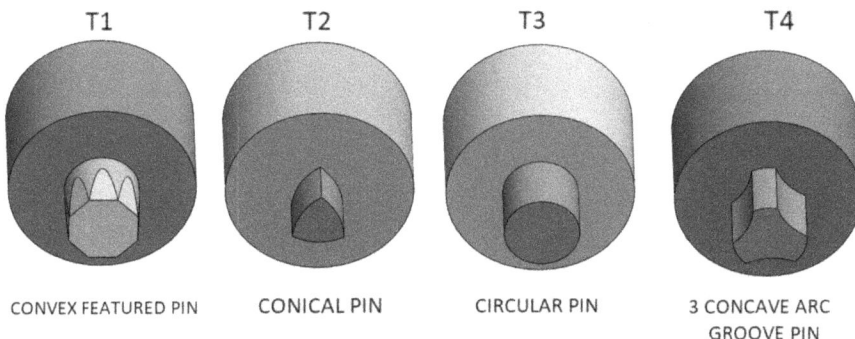

T1	T2	T3	T4
CONVEX FEATURED PIN	CONICAL PIN	CIRCULAR PIN	3 CONCAVE ARC GROOVE PIN

FIGURE 7.7 Schematic of the different tool pin geometry used for the FSAM.

that FSAM can be an improvement on the conventional manufacturing process by providing a finer and better microstructure in the manufactured products. Various magnesium alloys have been considered, including AZ31, Mg with Hydroxyapatite, WE43 Alloy, Al-Zn-Mg Alloy, AZ91 Alloys, and others. The microstructure of these alloys is studied to find the difference between the microstructures before and after the FSAM process. For this purpose, the Transmission Electron Microscope (TEM) is used. TEM microscopes can produce a highly magnified image by making use of a particle beam of electron.

A study was conducted to produce a composite of AZ31B Mg Alloy and Hydroxyapatite by Yee-Hsein Ho. Friction Stir Additive Manufacturing process was followed to form a composite of AZ31B and Hydroxyapatite (HA). Three different ratios of magnesium alloy to hydroxyapatite were taken: FSP Mg-20wt% HA, Mg-10wt% HA and Mg-5wt%HA composites. After the evaluation of the three structures under the Transmission Electron Microscope it was observed that the untreated AZ31 Alloy had a large grain size of up to 7μm, whereas after the FSAM process had been carried out the grain size of the structure had significantly reduced to 3.0 μm average. This reduction in grain size directly contributes to the increased durability of the material to stress (Ho et al., 2020). Similarly, another Alloy WE43 shows various changes in the microstructure grain size at different locations. This study was done by S. Palanivel et al. to develop a structurally efficient stack of Mg alloy WE43 so that it can be used for various real-world applications (Palanivel et al., 2015b). The various changes in the microstructure were studied under the SEM and TEM. The stacking pattern during the FSAM of WE43 Mg alloy is represented in the Figure 7.8. The grain size at different regions is represented in the Table 7.2.

A comparison was made between the base material and the stacked material that was produced by the process of FSAM. The BM was used to wrought operation and involved the use of rolling. It was detected that the BM consisted of the formation of intermetallic which mainly constituted of Mg-3.5%Y-2%ND-0.01%Zr. The formation of this intermetallic lead to decreased resistance to dislocations due to their large grain

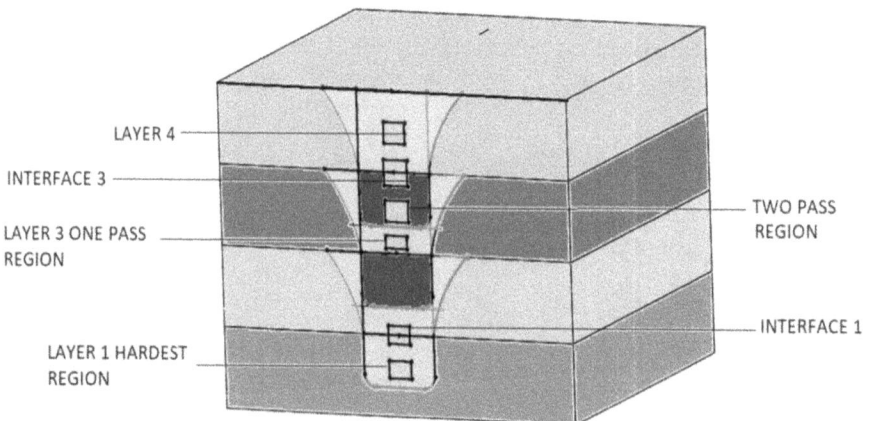

FIGURE 7.8 Stacked WE43 Magnesium Alloy. The green color represents a single pass region and the red color represent double pass region.

TABLE 7.2
Variation of Grain Size WE43 Mg Alloy by FSAM

Grain Location	Average Grain Size (µm)	Max Grain Size (µm)	Min Grain Size (µm)	Standard Deviation (µm)	LAGB HAGB (µm)
Layer 4	0.70	5.5	0.28	0.68	0.13
Interface 3	0.98	6.2	0.26	0.83	0.30
Layer 1	0.84	3.8	0.29	0.54	0.15
TMAZ	1.10	14.9	0.30	1.04	0.47

size of about 500 nm to 4 µm. These intermetallics have also been observed to consume a significant amount of yttrium atoms, because of which the full potential of the alloying elements is not utilized properly. These intermetallic compounds also cannot be removed by solutioning the mixture, as the temperature required to dissolve the intermetallic compounds is very high, i.e., more than 620 degrees. So, in order to achieve greater strength a higher number of dislocations is required, making the structure more inhomogeneous, in turn affecting the ductility of the metal. Thus, it was also established that a material produced through wrought operation can have higher strength at the expense of the ductility of the material, and that both cannot be achieved at the same time.

Thus, in order to overcome different shortcomings in case of FSP, the FSAM process was used. The experiment involved taking two different tool rotational speeds. One set of parameters was 1400 rev/min with a traverse speed of 102 mm/min (Palanivel et al., 2015b), and set of parameters was 800 rev/min, with a traverse speed of 102 mm/min. After the experiment, it was observed that the tool with a higher rotational speed was not able to generate enough strain rate, and there was formation of intermetallic compound found at the interface of the different layers, as the interfaces of the stack alloys were dominated by the action of the tool only, whereas there was fine grain structure observed at the area, which was dominated by the action of the tool shoulder. Whereas in case of the 800 rev/min, enough strain rate and temperature were provided to the material that there was no formation of intermetallic compounds, making the lower heat input parameter favorable for the FSAM process. At the highest tool rotational speed (1400rev/min), there was also the formation of joint lines and micro voids leading to the weaker nature of the produced stack of WE43 alloy. The build at 800 rev/min saw no such micro voids or joint line, uniform fine grain structure of 600 nm was formed at the non-interfacial region, and uniform grain structure of 2 µm were formed at the interfacial regions between the different layers of the WE43 alloy, making the FSAM process an ideal process over wrought operation to produce parts that have better microstructure and strength.

Just like the above-mentioned magnesium alloys, the reduction in microstructure size was further confirmed by an experiment done (Mukhopadhyay et al., 2020). They used Al powder on AA1060 substrate. It was observed that there was good intermixing between the layers of deposited Al powder and the AA1060 substrate. One point to be noted that there was no difference in microstructure throughout the traverse direction, which is generally seen with a tool pin involved friction stir additive manufacturing and all the zones (Nugget Zone, Thermo-mechanical affected zone, Heat affected Zone) being subjected to different stresses. The grain size of the deposited layer was 14.784 μm, which was found to be significantly smaller than the average grain size of the substrate (which was 167.739 μm). As a result of the Dynamic recrystallization of material due to the plastic deformation of the powder material, fine grains were produced (Mukhopadhyay et al., 2020). Dylan T. Peterson, David Garcia Hang Z. Yu and R. and Joey Griffiths also proved the refinement in grain structure of the previously used elongated microstructure in the AA7075 when they did an experiment to fill a hole in AA7075 with the help of the FSAM process. They observed the grain refinement up to 3.4 ± 0.7 μm in the deposited material inside the hole of 6.5 mm that was to be repaired (Griffiths et al., 2019). The schematic of the hole filling operation is represented in the Figure 7.9.

Sahu et al. (2020) observed the BSE microstructural mapping as shown in the Figure 7.10. The specimen's elemental composition at 6% further added Al alloy is shown in Figure 7.10. The α-Mg alloy and eutectic phase are exactly distributed as seen in the colored and grey figure, respectively. The mapping showed that the Al is in the grey zone and the α-Mg is in the black zone. The mapping clearly demonstrated proper heat generation from the FSSA process, appropriate diffusion of the alloyed Al with Mg matrix due to proper stirring action and plastic deformation. It was determined from the mapping that orange color reflects Mg distribution and the green color represents Al. According to the distribution pattern, Mg serves as a matrix, and Al IMCs are distributed evenly on Mg. IMCs are distributed uniformly in a layer-by-layer pattern throughout the specimen of Mg matrix. The IMCs everywhere to Mg were strongly and uniformly bonded like a honeycomb structure, which improves the mechanical characteristics and maximizes tensile strength.

The effect of alloying powder of Mg abstract has some microstructural effect. The detail analysis of Sahu et al. (2020) is again discussed as follows. It can be challenging to distinguish grains of Mg and Al because of similar looks because the features of Mg grain and the additional Al grain structures are so similar. The differentiation of grain boundaries is less visible due to the homogeneous dispersion of the Al alloy. However, after chemical etching, the separated Al alloy appears as a black mark, as seen in Figure 7.11. Due to the ease of externally introduced Al particles nucleating,

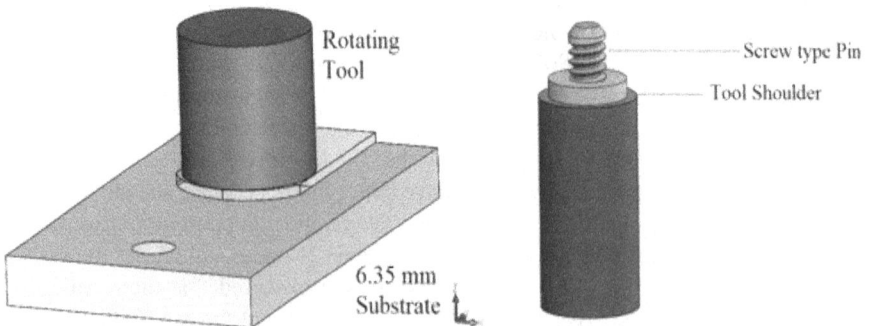

FIGURE 7.9   Hole filling operation using FSAM.

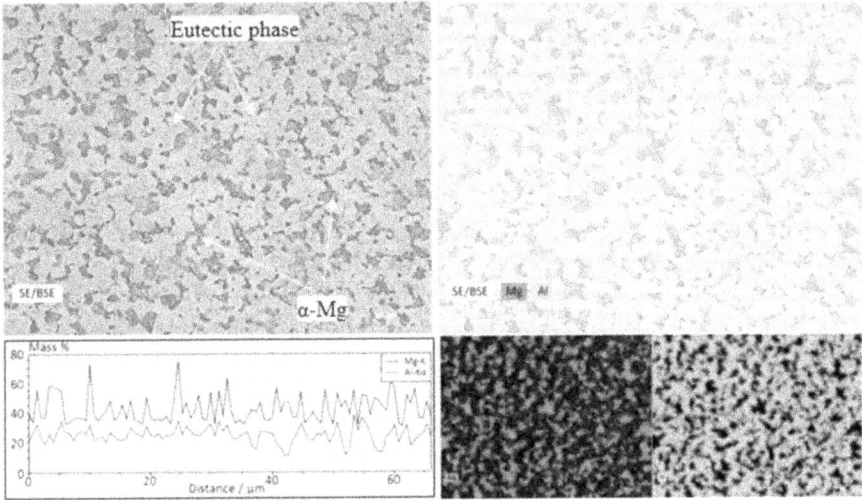

FIGURE 7.10 BSE elemental distribution FSAM sample with 6% Al.

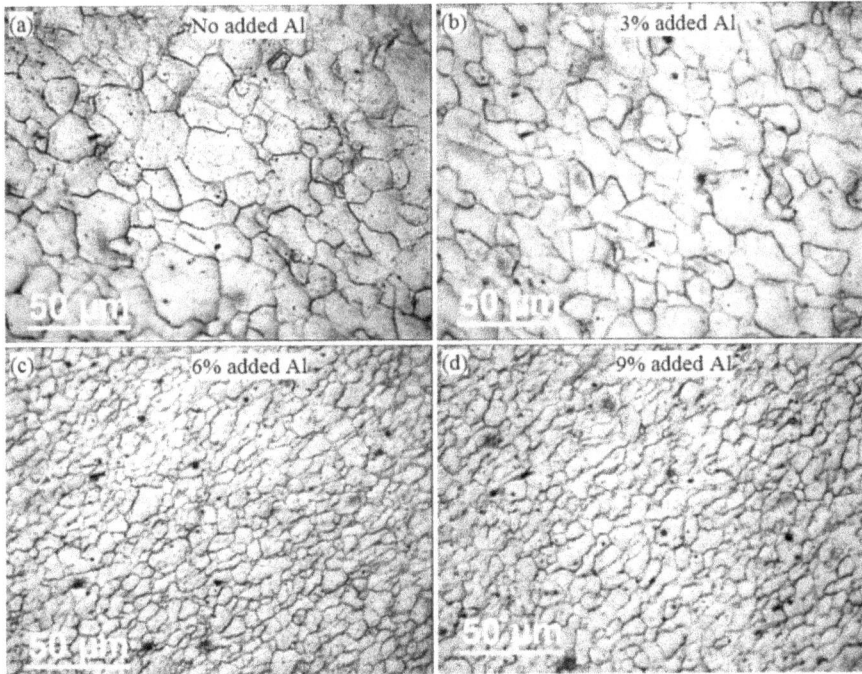

FIGURE 7.11 NZ grain size variation with Al%, (a) FSP sample, non-alloyed Al, (b) 3% (c) 6% and (d) 9% added Al.

unlike Al%, has an impact on grain size difference throughout the process. The additional Al particle serves as a foundation for facile nucleation and grain formation during recrystallization. In Figure 7.11a–d, the FSP sample with and without external Al addition at 3, 6, and 9% to the NZ are depicted using the same process conditions. The grain size for the FSP sample at NZ was found to be 22 μm and roughly 66% fine to BM. Additionally, the NZ grain got finer as the Al% went up. The FSP section receives more nucleation bases as the Al% rises, which leads to the creation of more new grains in the same amount of treated surface, which reduces the average grain size. As a result, the grain size for the 3, 6, and 9% Al alloyed samples, respectively, is 14, 8, and 7 μm. Due to the aforementioned rationale, it was found that the alloying specimens produce smaller grain than the unalloyed sample. The 6% alloyed NZ grains are roughly three times smaller than the FSP specimens. But it was found that the smallest grain (7 μm) related to 9% Al alloying. The strength of the sample should increase as the grain size decreases, according to grain analysis. However, with 9% Al, the FSSA sample's strength is reduced as a result of the increased percentage of externally added Al particles aggregating and agglomerating, which worsens its tensile properties compared to the 6% Al scenario (Table 7.3).

TABLE 7.3
Representation of the Microstructural Variation by Different Researcher

Authors Name	Alloy Used	Microstructure of Base Alloy	Microstructure of the Alloy after FSAM	Conclusion
Griffiths et al. (2019)	Al Alloy 7075	Microstructure of the base metal was found to be in the range of 100's of micron, which is usually the average size of grains in rolled Al 7075 Alloy	After the hole filling operation there was significant decrease in the size of the microstructure to 3μm	FSAM was proved to be effective for the hole filling operation, as the deposited material and the side walls of the groove had the same refined grain size.
Srivastava and Rathee (2020)	Al alloy 5059/Sic	The base alloy of aluminum had coarse grain structure after the rolling operation	FSAM successfully managed to refine the grain structure to about 1 μm micron	The final composite of al alloy and silicon exhibited superior quality of grains, thereby increasing its strength.
Palanivel et al. (2015a)	WE 43 Mg Alloy.	Microstructure of the base metal was seen to exhibit intermetallics with grain range up to 500 μm	After the FSAM the grain reduction takes place up to 1.5 μm	It was concluded that process parameters affected the grain size of the resultant alloy, especially the tool rotation speed.

7.5 MECHANICAL PROPERTIES OF MAGNESIUM ALLOYS PRODUCED BY FSAM

Mechanical properties are the physical properties that a material possesses as a result of the various forces that have been applied to it. Examples of mechanical properties are tensile strength, hardness, fatigue limit, elongation, and modulus of elasticity. Mechanical properties analysis must be necessary to get the material properties after the FSAM process and can be compared with the initial base metal to get a conclusions.

7.5.1 HARDNESS ANALYSIS

To evaluate the properties of a material, a method is used that is called hardness test, which finds out the strength, ductility, and wear resistance of a material and helps to determine whether the material can be used for the desired application or not. The most common type of hardness test used by the authors was the Vickers Micro-Hardness test, which makes use of a diamond pyramid indenter. After the application of load the length of the diagonal indent gives the hardness information of a material (Fatemi, 2000). As has already been discussed, with the changes in the process, parameters, and tool geometry there can be a significant change in the microstructure of a material. In 1951, E. O. Hall wrote three papers in which he mentioned the relationship between the hardness and the grain size of a material. In 1953, N. J. Petch of the University Leeds, England published a paper after doing a series of experiments and confirming the relationship put forward by E. O. Hall. This was then developed into an equation.

$$H = H_0 + k_H d^{-1/2}$$

H denotes the hardness of the material, diameter of the average grain size is represented by the letter 'd', H_0 and K_H are the Hall Petch constants. This relation states that the Hardness and strength of a material is inversely proportional to the grain size. It was also observed by the authors that the strength of the material depended upon dislocations, solute element grain boundaries second phase to block the dislocations, and that the nature of the microstructure was highly dependent on the processing and welding parameters.

In an experiment conducted by Palanivel et al. (2015a) to produce a structurally improved stack of WE43 Alloy it was found that the component produced through FSAM gives enhanced strength. It was found that the structure obtained after the parameter of 1400 rev/min and 102 mm/sec traverse speed the hardness of the material was 115 HV, which further improved to 135 HV after the ageing process was carried out. It should be noted that the hardness result obtained by FSAM under the given process parameters was much better than the hardness obtained from the conventional heat treatment technique that have 110–120 HV. Yee Sein Ho and others found the hardness of AZ31B Mg/Ha alloy to improve from 78 HV to 95 HV after the composite was prepared following an operation of FSAM. Similarly, Srivastava et al. (2019), observed an improvement of hardness from 85 HV to 140 HV peak

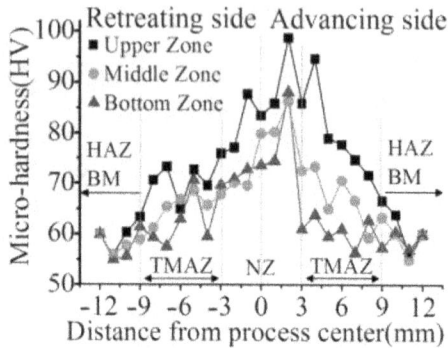

FIGURE 7.12   Micro hardness variation at different zones.

hardness of Al alloy. They established that this was due to the fact that Al5059 is a non-heat treatable alloy, so there was no occurrence of softening and further the grain refinement led to the increase in the resistance of the material to deformation as put forward by Hall and Petch.

However, another experiment conducted by Garcia et al. (2020) astonishingly showed a decrease in the hardness of the deposited AA7075 while repairing a hole. The hardness of the deposit is 164 HV, but after the Friction Stir Additive based Deposition was carried out, it was observed that the average hardness of the repaired hole had reduced to 142 HV Garcia et al. (2020). Like other experiments, Friction Stir Deposition technique was used for the purpose of repairing the holes, As expected, the authors found a decrease in the grain size of the deposited AA7075 from several hundred microns to a few microns. The variation in grain size from the feed material and the deposited material usually means that there should be an increase in the strength and hardness of the material. However, in this case the temperature of the FSAM operation led the temperature to reach the 80–90% of the melting temperature of Al which is also the solutionizing temperature of the AA7075 alloy. Even though the material was exposed to such high temperatures for a short period of time, it was concluded that the dwell time during reparation could have resulted in the dissolution of a significant amount of the precipitates into the lattice. This led to considerably less precipitation hardening resulting in the finer microstructure of the deposited material. Sahu et al. (2020) observed variation of the hardness at a different zone of the friction stir additive Mg alloy as represented in the Figure 7.12.

7.5.2 TENSILE STRENGTH ANALYSIS

Tensile testing is a type of destructive test, which is carried out to find the information about the ultimate tensile strength (UTS), yield strength (YS) and ductility of the considered material. This test gives the amount of force that is required for the elongation to give the YS and stretchable up to the limits of UTS of the specimen, and at last its break. Thus, a tensile test was carried out by many researchers to get the improvement in the FSAM processed samples compared with the base material (Table 7.4).

TABLE 7.4

Representation of Different Researcher Outcomes

Author Name	Alloy Used	Hardness Before the FSAM Process	Hardness Values Obtained after FSAM	Conclusion
Ho et al. (2020)	AZ31B Mg Alloy, Hydroxyapatite	The initial hardness of the base metal was found to be 78 HV	The final hardness obtained after the FSAM Process was 98 to 105 HV	The increase in hardness was proved to be better for the purpose of using it as a bio implant for bone replacement.
Palanivel et al. (2015a)	WE43 Mg Alloy	The hardness for the base WE43 alloy was found to be	After the FSAM process, the hardness was significantly improved to 110 HV	Due to the increase in the hardness value of the specimen
Griffiths et al. (2019)	Al alloy 7075	Initial hardness before the FSAM was observed to be 160 HV	After the FSAM the hardness of the alloy was found to decrease to 140 HV	The authors concluded that the hardness of the alloy decreased due to the high rotational speed of the tool that resulted in the alloy reaching its near melting point

Some researchers use powder mixer of 40 atomic percentage of graphite with a balanced Al and then filled the groove of the Al plate with this powder mixture. They found that the UTS of the sample is 147.5 MPa, which was much superior to base Al. Also, the elongation of the composite material was found to be 26% more than that of the base Al alloy. Some other researchers conducted another experiment in which they punched holes in the bottom of the Al-Mg consumable metal rod and then filled these holes with zinc powder. The resultant peak strength was found to be 213.83 MPa at an 8-hour aging time. However, 4-hour aging time led to the increase in size of the crystal grains resulting in the formation of non- homogeneous precipitates, thereby degrading the performance. Whereas the 8 hours aging process led to the precipitates of the intermetallic compounds containing zinc, therefore improving the strength (Table 7.5).

Mukhopadhyay et al. (2020), experimented on Al powder deposit-based FSAM process and showed significant improvements in UTS and elongation percentage of sample. This was shown by the initial elastic region followed by plastic deformation up to the fracture point. The Authors found the UTS to be 112.5 MPa and percentage elongation to be 24.663%. Also, it was observed that the tensile yield strength of the deposited material via FSAM was 78 MPa, a significant improvement to the AA1060

TABLE 7.5
Summary of FSAM of Few Mg and Al Alloys

Authors Name	Experiment Done	Alloy Used	Process Parameters	Initial Microstructure	Final Microstructure	Initial Hardness	Final Hardness	Remarks
Griffiths et al. (2019)	Repair of Al 7075 Alloy by the application of FSAM Process	Al Alloy 7075	Tool Rotation Speed- 400 RPM, Traverse Speed- 0.32mm/s	Varies in 100 of μm	0.7–3 μm	160 HV (Average)	Peak- 172 HV Average- 145 HV	Increase in Grain refinement but slight decrease in hardness.
Srivastava and Rathee (2020)	Fabrication of Al 5059 and Silicon Carbide	Al Alloy and Silicon Carbide	Tool Rotation Speed- 450 RPM, Traverse Speed- 63 mm/min	3–5 μm	0.5–1 μm	85 HV (Average)	140 HV	Increase in Grain refinement and also increase in Hardness.
Palanivel et al. (2015a)	To obtain High Structural Performance in Magnesium Alloy	WE43 Magnesium Alloy	Tool Rotation Speed- 800 RPM and 1400 RPM, Traverse Speed- 102 mm/min	4 μm	1.5–2.5 μm	110 HV	115 HV and 135 HV after aging	Increase in Grain refinement and Microhardness with UTS up to 400 Mpa
Ho et al. (2020)	To obtain better corrosion resistance in Magnesium Alloy for application in the bio implant sector	AZ31B Magnesium Alloy, Hydroxyapatite	Tool Rotation Speed- 900 RPM, Tool Traverse Speed- 30 mm/min	7.7–1.6 μm	2.2–0.25 μm	78 HV	95 HV	Increase in grain refinement and hardness and increase in corrosion resistance.

and AA1060 H12. The author determined that the improvement in the UTS was for finer grain and dynamic recrystallization of the deposit. The improvement in percentage elongation indicating super plasticity in the material was attributed to the thermomechanical effect of FSAM process which led to dynamic recrystallization.

He et al. (2020) experimented on to build a multi-layered structure using Al–Zn–Mg alloy involving FSAM operation found that significant increase in the hardness for up to 60 days after that decrease from maximum hardness. The experiment result showed the UTS of 483.1 MPa and 20.2% elongation in traverse direction. Whereas the average UTS was found to be 294.0 MPa with an elongation of 13% after 5 days of natural aging. This was an improvement of UTS and elongation by 61% and 65% respectively as compared to BM.

The mechanical qualities of the FSAM product are also influenced by the percentage of alloying elements. Some of the data given here were acquired from Sahu et al. (2020). The stress/strain flow curve is shown in Figure 7.13a for BM and FSP sample. FSP sample's effectiveness is lower and equals to 79% of BM and which is lower because of less ductility, as seen by Figure 7.13a. So, the Al particles are introduced to the processing area in concentrations of 3, 6, and 9% to increase ductility and tensile strength. According to the overall tensile properties, the UTS, YS, and elongation % of all increased in comparison to the BM at 6% Al. These improvements are 95%, 97.5%, and 77%, respectively. However, for sample with 6% Al have, the UTS, YS, and elongation % are 118%, 134%, and almost 200% of the non-alloying treated base metal. Additionally, it was found that specimens with a 6% Al alloy have superior mechanical qualities to those with a 9% Al alloy. A higher Al to Mg base ratio produces the $Mg_{17}Al_{12}$ compound, which has negative effects on mechanical characteristics.

The UTS, YS, and elongation %, are shown in Figure 7.13b. In comparison to an alloying sample, the 0% alloying scenario provides less effective tensile properties. Tensile characteristics at 3% Al alloying are lower because of inadequate IMC alloying material. However, the tensile parameters mentioned above improve with 6% Al alloying for appropriate and uneven dispersal of alloy. But for accumulation of the dispersed particle at 9% Al, the same property reduces. The ultrafine dispersion particle and reinforcing Mg-Al IMCs both contribute to alloying's greatest strength.

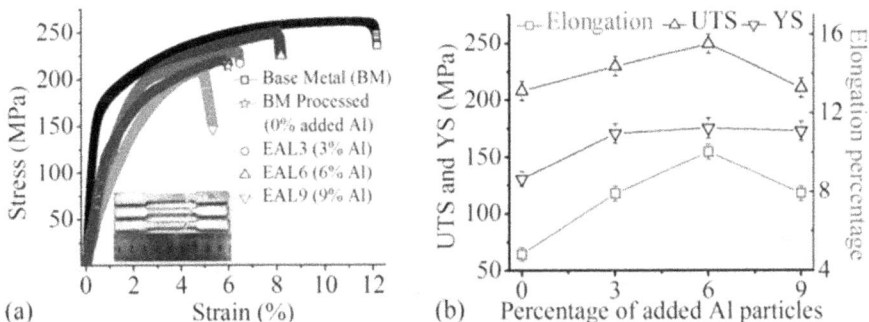

FIGURE 7.13   (a) Stress vs. strain graph for different samples (b) UTS, YS and elongation % at different Al%.

According to this research, the selective alloyed part's strengthening effect results from a successful load transmission from Mg matrix to alloying Al via reinforcing IMCs. It should be mentioned that the load transmission is a result of assured IMCs at a specific size.

7.6   CONCLUSIONS

- From the above preliminary approach, it can be said that the FSAM process is a modern, fast, size independent advanced metal AM technique. This method can be effectively used to produce fully dense and accurately shaped wrought components economically. From the previous research work it was found that there was a significant improvement in microstructure, hardness, and tensile properties of materials after the FSAM process was implied. It was observed that the process parameters played an important role in deciding the resultant mechanical properties of the materials.
- The microstructure of the material matrix plays a vital role in the resultant hardness and the tensile strength of the material. Tool Geometry is another parameter that should be explored and experimented upon so as to find the most suitable one for the given job. A finer microstructure does not always result in betterment of FSAM produced components that exhibit superior strength, which results from highly refined grain sizes ranging 2–7 nm.
- Corrosion resistance of a material like AZ31B can be increased by the application of FSAM, making it suitable to be used in bio implants. It has been concluded that FSAM is an effective technique which can be employed for the purpose of repairing holes as in the case of AA7075. Even wide grooves can be filled effectively with this process.
- The speed of the rotating tool can highly influence the final structural build, as in the case of hole repairing of AA7075 in which 15% of the hardness was lost due to the thermo-mechanical effect of the tool. This FSAM technique is a significant improvement to the conventional manufacturing techniques and should be employed in the modern-day manufacturing sector for maximum efficiency and productivity.

REFERENCES

Ahluwalia S. and Mahto R. V., 2018. Additive manufacturing based innovation, small firms, customer involvement and crowd-funding: from co-creation to co-financing, *Transl. Mater. Res.* 5, 026001.

Ali A., Ahmad U. and Akhtar J., 2020. 3D printing in pharmaceutical sector: An Overview. https://doi.org/10.5772/intechopen.90738

Chua Z. Y., Ahn H., Moon S. K., 2017. Process monitoring and inspection systems in metal additive manufacturing: Status and applications, *International Journal of Precision Engineering and Manufacturing-Green Technology* 4(2), 235–245.

Derazkola H. A., Khodabakhshi F., Simchi A., 2020. Evaluation of a polymer-steel laminated sheet composite structure produced by friction stir additive manufacturing (FSAM) technology, *Polymer Testing* 90, 106690.

Fatemi A., 2000. Mechanical properties and testing of metallic materials. https://doi.org/10.1002/14356007.b01_10

Gao H., Li H., 2021. Friction additive manufacturing technology: A state-of-the-art survey, *Advances in Mechanical Engineering*, 13(7), 1–29. https://doi.org/10.1177/16878140211034431

Garcia D., Hartley W. D., Rauch H. A., Griffiths R. J., Wang R., Zhenyu, Kong J., Zhu Y., Yu H. Z., 2020. In Situ Investigation into Temperature Evolution and Heat Generation during Additive Friction Stir Deposition: A Comparative Study of Cu and Al-Mg-Si, https://www.sciencedirect.com/science/article/pii/S2214860420307582

Griffiths R. J., Petersen D. T., Garcia D., Yu H. Z., 2019. Additive friction stir-enabled solid-state additive manufacturing for the repair of 7075 aluminum alloy, *Applied Science*, 9, 3486. https://doi.org/10.3390/app9173486

He C., Li Y., Zhang Z., Wei J., Zhao X., 2020. Investigation on microstructural evolution and property variation along building direction in friction stir additive manufactured Al–Zn–Mg alloy, *Materials Science & Engineering A* 777, 139035, https://doi.org/10.1016/j.msea.2020.139035

Ho Y. H., Joshi S. S., Wu T. C. , Hung C. M., Ho N. J., Dahotre N. B., 2020. In-vitro bio-corrosion behavior of friction stir additively manufactured AZ31B magnesium alloy-hydroxyapatite composites, *Materials Science & Engineering C* 109, 110632. https://doi.org/10.1016/j.msec.2020.110632

Honarvar F., Varvani-Farahani A., 2020. A review of ultrasonic testing applications in additive manufacturing: Defect evaluation, material characterization, and process control, *Ultrasonics* 108, 106227.

Horvath D., Szabo R. Z., 2019, Driving forces and barriers of Industry 4.0: Do multinational and small and medium-sized companies have equal opportunities? *Technological Forecasting & Social Change* 146, 119–132.

Jahangir Md N., Mamun M. A. H., Sealy M. P., 2018, A Review of Additive Manufacturing of Magnesium Alloys, *AIP Conference Proceedings*, 030026. https://doi.org/10.1063/1.5044305

Jain R., Pal S. K., Singh S. B., 2018. Thermomechanical Simulation of Friction Stir Welding Process Using Lagrangian Method, Simulations for Design and Manufacturing, Lecture Notes on Multidisciplinary *Industrial Engineering*, https://doi.org/10.1007/978-981-10-8518-5_4

Lee J. Y., An J., Chua C. K., 2017. Fundamentals and applications of 3D printing for novel materials, *Applied Materials Today* 7, 120–133.

Li W. Y., Li J. F., Zhang Z. H., Gao D. L., Chao Y. J., 2013. Metal flow during friction stir welding of 7075-T651 aluminum alloy, *Experimental Mechanics*, 53:1573–1582. https://doi.org/10.1007/s11340-013-9760-3

Mukhopadhyay A., Saha P., 2020. Mechanical and microstructural characterization of aluminium powder deposit made by friction stir based additive manufacturing, *Journal of Materials Processing Tech.* 281, 116648, https://doi.org/10.1016/j.jmatprotec.2020.116648

Palanivel S., Nelaturu P., Glass B., Mishra R. S., 2015a. Friction stir additive manufacturing for high structural performance through microstructural control in an Mg based WE43 alloy, *Materials and Design* 65, 934–952. https://doi.org/10.1016/j.matdes.2014.09.082

Palanivel S., And Sidhar H. Mishra R. S., 2015b. Friction stir additive manufacturing: Route to high structural performance, *JOM*, 67(3). https://doi.org/10.1007/s11837-014-1271-x

Pareek M., Polar A., Rumiche F., Indacochea J. E., 2007. Metallurgical evaluation of AZ31B-H24 magnesium alloy friction stir welds, *JMEPEG*, 16, 655–662. https://doi.org/10.1007/s11665-007-9084-5

Prakash K. S., Nancharaih T., SubbaRao V. V., 2018. Additive manufacturing techniques in manufacturing -An overview, *Materials Today: Proceedings* 5, 3873–3882.

Puleo S. M., 2016. Additive friction stir manufacturing of 7055 aluminum alloy, University of New Orleans, https://scholarworks.uno.edu/honors_theses/75

Sahu P. K., Das J., Chen G., Liu Q., Pal S., Zeng S., Shi Q., 2020. Friction stir selective alloy-ing of different Al% particulate reinforced to AZ31 Mg for enhanced mechanical and metallurgical properties, *Materials Science & Engineering A* 774, 138889, https://doi.org/10.1016/j.msea.2019.138889

Shunmugam M. S., Kanthababu M., 2018. Advances in Additive Manufacturing and Joining, *Proceedings of AIMTDR*, https://link.springer.com/book/10.1007/978-981-32-9433-2

Srivastava M., Rathee S., 2020, Microstructural and microhardness study on fabrication of Al 5059/SiC composite component via a novel route of friction stir additive manufacturing, *Materials Today: Proceedings*. https://doi.org/10.1016/j.matpr.2020.07.137

Srivastava M., Rathee S., Maheshwari S., Siddiquee A. N., Kundra T. K., 2019. A review on recent progress in solid state friction based metal additive manufacturing: Friction stir additive techniques, *Critical Reviews In Solid State and Materials Sciences*. https://doi.org/10.1080/10408436.2018.1490250

Stelt A. A., Bor T. C., Geijselaers H. J. M., Akkerman R., Boogaard A. H., 2013. Cladding of Advanced Al Alloys Employing Friction Stir Welding, Key Engineering Materials Vols 554-557, 1014-1021 Trans Tech Publications, Switzerland doi:10.4028/www.scientific.net/KEM.554-557.1014

Wolff S. J., Lin S., Faierson E. J., Liu W. K., Wagner G. J., Cao J., 2017. A framework to link localized cooling and properties of directed energy deposition (DED)-processed Ti-6Al-4V, *Acta Materialia* 132, 106–117.

Yu C., Xu X., Lu Y., 2015. Computer-integrated manufacturing, cyber-physical systems and cloud manufacturing – Concepts and relationships, *Manufacturing Letters* 6, 5–9.

8 Analysis and Improvement Performance of Manufacturing in Friction Stir Welding

Sharda Pratap Shrivas

Department of Mechanical Engineering, BIT Durg, Chhattisgarh, India

G. K. Agrawal

Department of Mechanical Engineering, GEC, Bilaspur, India

Shubhrata Nagpal

Department of Mechanical Engineering, BIT Durg, Chhattisgarh, India

Shailesh Dewangan

Chouksey Engineering College, Bilaspur, Chhattisgarh, India

CONTENTS

DOI: 10.1201/9781003346623-8

119

8.1   INTRODUCTION

Evolution of Friction Stir Welding (FSW) is the latest traditional and advanced process for power welding to join two similar or different aluminum alloys. FSW avoids many of the challenges brought on by the change in the material's condition that frequently occur in conventional welding methods because it depends on heat source; this source of heat can be produced with the help of friction and plastic work (Burek et al., 2017). The process of welding is dependent on various parameters, such as tool material, rotation speed, and heat generation for production of a quality product. When exposed to high temperatures, the material and tool rotation speed should have good shear strength, wear resistance, and oxidation resistance (Azadi Chegeni and Kapranos, 2018). The FSW welding process gives the best results by creating friction stir welding between the two workpieces, one of the cutting-edge, fastest-growing metal joining methods turns the contact pin of the FSW tool at the workpiece into a welded junction. (Rajeesh, Balamurugan, and Balachandar, 2018).

Microstructure's properties of welded joints are deviate for different mechanical structure change, which can be seen with the help of a scanning electronic microscope (Zhang et al., 2023; Chen et al., 2023). Several plastic deformation approaches are also affected; the mechanical properties also change the orientation of atomic structure with slip and twining deformation (Raj and Biswas, 2023; Sathish et al., 2022). Friction Stir Welding is also used for a magnetic field for using ultra high speed for rotation of the stirring head (Chen et al., 2023). The variety of work on the stir welding process is discussed by special analysis on past performance. There is a lack of researchers finding the welding strength and Charpy strength, so this paper analysis shows the optimized welding strength and Charpy test with a decent setting of machining parameters.

8.2   MATERIALS AND METHOD

In the FWS process, there are two different sorts of factors: controllable parameters that can be modified or controlled, and uncontrollable parameters that may have an impact on welding but that cannot be controlled, such as temperature, vibrations, human error, etc. Welding speed, tool rotation speed, tool tilt angle, tool pin profile, downward force, shoulder and pin diameter, etc., are all controllable characteristics. Selected machining parameters are temperature, holding time for pre heat treatment and tool rotation speed, these parameter factors and their levels are shown in Table 8.1. According to investigation, temperature holding time and speed to tool rotational are the major effecting parameters for the strength of welding joints.

TABLE 8.1
Factors and Their Levels

Factors	Unit	Level 1	Level 2	Level 3
Temperature	°C	200	250	300
Holding time	Min.	40	50	60
Speed of tool rotation	rpm	600	800	1000

TABLE 8.2
Design of Experiment

Run no.	Temperature (in °C)	Holding Time (in minute)	Speed of Tool (rpm)
1	200	40	600
2	200	50	800
3	200	60	1000
4	250	40	800
5	250	50	1000
6	250	60	600
7	300	40	1000
8	300	50	600
9	300	60	800

MINITAB is the software used for design of experiment (DOE) techniques. In this case, the three factors with three level gives 27 experiments in full factorial design, but it can be compressed by using L_9 orthogonal array design (OAD). L_9 orthogonal array design used for welding the specimen that provides nine welding parameters design is shown in Table 8.2.

8.2.1 PROCEDURE OF SPECIMEN

In this paper, the workpiece is selected as aluminum alloy with the dimension of 4 mm thick sheet plate. The parts are initially created in the correct size, with dimensions of $200 \times 100 \times 4$ mm, using a power hacksaw. These workpieces are passed into the oven as per design of experiment, then the workpieces are static in the fixture. The tool with the desired dimensions was prepared, which is 16 mm shoulder diameter, and is used with 3.5 mm pin length. The welding was carried out in accordance with the variation in tool rotation speed specified in the experiment design after feeding the rotating tool along the centerline. On the nine samples recommended by the L_9 orthogonal array approach, a tensile test was conducted. For measurement of tensile, test specimens were prepared with standard dimensions. This dimension is shown in Figure 8.1.

A standardized high strain-rate test that measures the amount of energy absorbed by a material during fracture is the Charpy impact test, sometimes referred to as the Charpy V-notch test. Charpy test specimens are made in accordance with ASTM E23

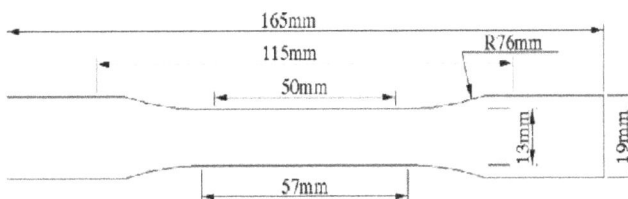

FIGURE 8.1   Tensile Test Specimen Dimension.

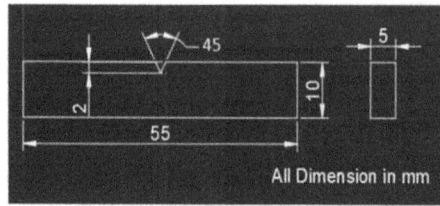

FIGURE 8.2   Charpy Specimen Dimension.

TABLE 8.3
Responses of Tensile and Charpy Strength

S. No.	T (°C)	HT (Sec.)	S (rpm)	Ts (MPa)	Ct (Joule)
1	200	40	600	67	21
2	200	50	800	88	42
3	200	60	1000	64	27
4	250	40	800	86	28
5	250	50	1000	68	53
6	250	60	600	53	34
7	300	40	1000	68	27
8	300	50	600	62	42
9	300	60	800	73	22

Where as,
T = temperature, Ts = Tensile Strength in MPa, HT = Holding Time, Ct = Charpy Strength in Joule, S = Tool rotation Speed

and tested using Charpy test equipment that is presented in Figure 8.2. The test criteria of drop angle is set as 160°, and impact energy was set as 368 J with pendulum direction, and velocity of hammer is 5.2 m/s for striking process.

To assess the tensile strength of the joints, nine tensile and Charpy test specimens were created in accordance with ASTM E8 and E23 standards from the American Society for Testing of Materials. The findings of the Charpy test, which was used to assess the tensile strength of the FSW joints, are shown in Table 8.3.

8.3 TENSILE AND CHARPY STRENGTH COMPARISON

Increasing the parameter strength will result in an increase, as can be observed in all comparison graphs. The comparison of the tensile test response for water- and air-cooling conditions is shown in Figure 8.3. It is evident from this graph that raising the parameter levels would likewise raise the strength factor.

Now prepare the graph of comparison of temperatures for the Charpy test response which is shown in the Figure 8.4. The graph is first increase and then decreases then again increase which is due to other parameter effect, but overall, the hardness of the material affects the strength of the weld.

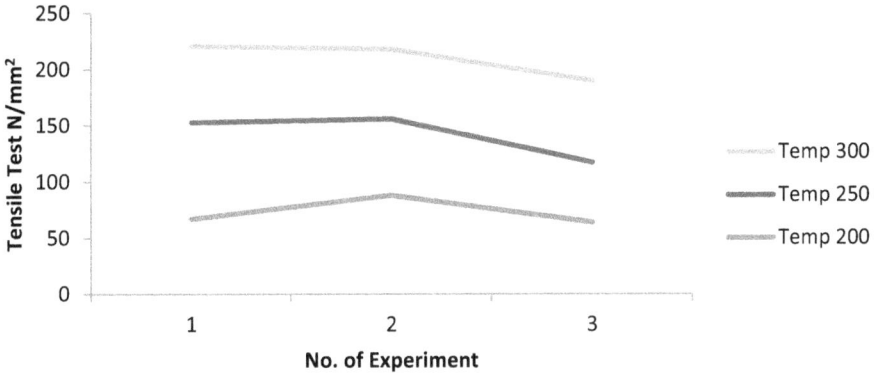

FIGURE 8.3   Comparison of tensile test between temperatures.

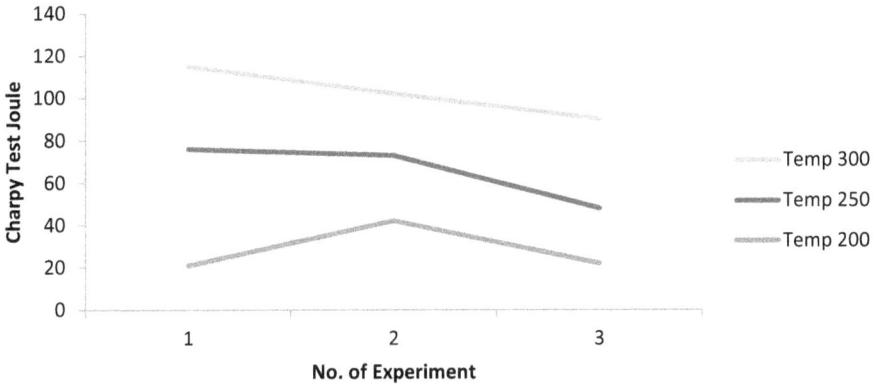

FIGURE 8.4   Comparison of Charpy test between temperatures.

8.4   ANOVA ANALYSIS

On the base of Analysis of Variance (ANOVA) result for stir welding responses tensile test and Charpy strength test results are presented in Tables 8.4 and 8.5. The factor graph for both sets of responses has been created to identify the important component for each output response. Analyses of Variance are tested for 5% significant level, where source of P value shows less than 0.05 is the significant factor.

ANOVA analysis for the tensile test and Charpy test are obtained, P value 0.015 and 0.013 respectively, for holding time. For both cases, holding time is the significant factor under 5% significant. Therefore, it is evident from the graph that the points or values of parameters with the peak positions on the graph can be used to determine the ideal welding parameter. The following are the factor levels for predictions: Temperature 200°C, Holding Time 40 minutes and Rotation Speed of tool is 800rpm for tensile test but there is deviation in parameters for Charpy test are Temperature 250°C, Holding Time 50 minutes and Rotation Speed of tool is 1000rpm, given the best result in these parameters settings. Figures 8.5 and 8.6 shown the main effect plot for tensile and Charpy test result respectively.

TABLE 8.4
ANOVA for Tensile Test

Source	DF	Seq SS	Adj SS	Adj MS	F	P
T	2	46.222	46.222	23.111	16.00	0.059
HT	2	194.889	194.889	97.444	67.46	0.015
S	2	750.889	750.889	375.444	259.92	0.004
Residual Error	2	2.889	2.889	1.444		
Total	8	994.889				

TABLE 8.5
ANOVA for Charpy Test

Source	DF	Seq SS	Adj SS	Adj MS	F	P
T	2	133.556	133.556	66.778	13.98	0.067
HT	2	742.889	742.889	371.444	77.74	0.013
S	2	38.889	38.889	19.444	4.07	0.197
Residual Error	2	9.556	9.556	4.778		
Total	8	924.889				

Main Effect Plot for Tensile Strength

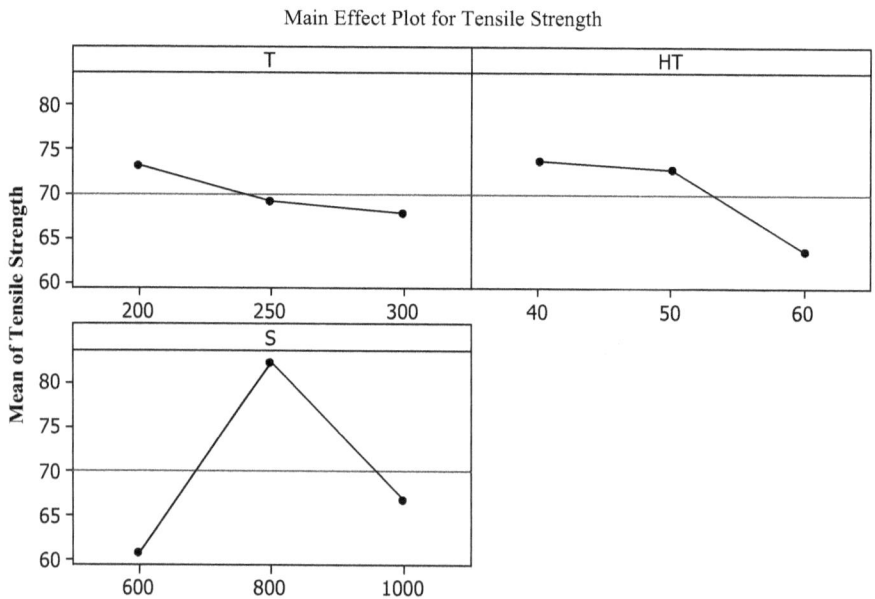

FIGURE 8.5   Main effect plot for tensile test.

Main Effect Plot for Charpy Test

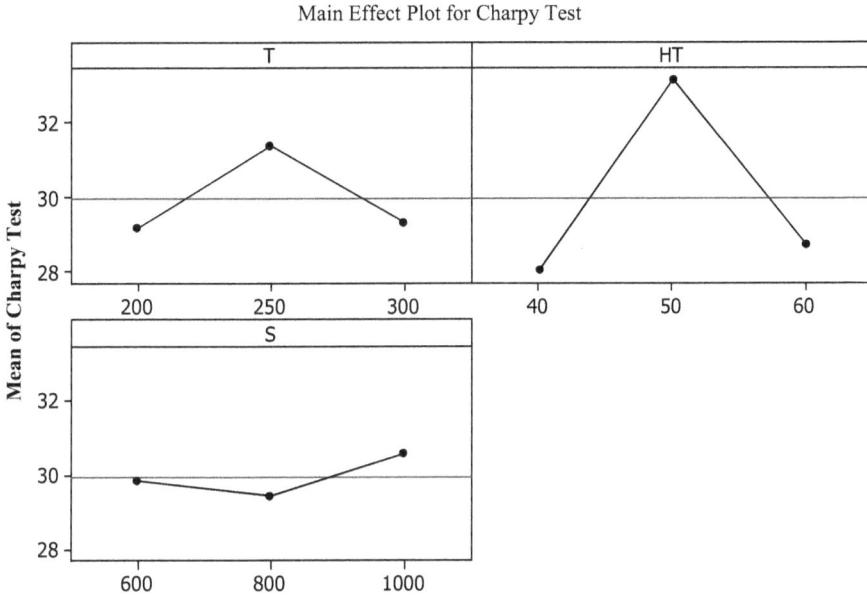

FIGURE 8.6 Main effect plot for Charpy test.

8.5 MICROSTRUCTURE ANALYSIS

In addition, the stir zone's characteristic appearance didn't find any crack in the specimen, which is found optimum by ANOVA analysis. Again, the optimum parameters are performed and check the strength of welding, which is verified by SEM image. Besides, the stir zone is dendritic, the weld microstructure is entirely austenite, and the region near the stir line is a columnar crystal grown circularly to the stir line to the center of the stir pool (see Figure 8.7a). Thus, the stir zone's characteristic appearance. There is a very noticeable coarse-grained area outside the stir line that is impacted by the welding thermomechanical cycle during the welding operation. The base material is only minimally impacted by heat because of the concentrated energy of the laser beam, which also causes the steel plate to melt and cool quickly. This results in a very small heat affected zone. The intense plastic deformation, material flow, and temperature cycle imposed during the welding process are responsible for the ensuing microstructures. Figure 8.7b displays a macrograph of the weld in cross-section.

The microstructure of specimen number three, which breaks from the neck during tensile testing, was observed. On both sides of the thermo-mechanically affected zone (TMAZ) zone, which has been covered, the grain size is larger than that of the Nanking safety zone (NSZ). Figures depict how the FSW zones with very fine grain size appear. Figure 8.8 depicts the consistent distribution of the plastic flow behavior of the different alloys as an oval shape. Additionally, it shows that both the NSZ and TMAZ have identical grain sizes but different geometries. Applying enough heat generation prevents piping and tunnel warming faults. The microstructure is clearly intended to produce a weld free of defects for the proper heat generation.

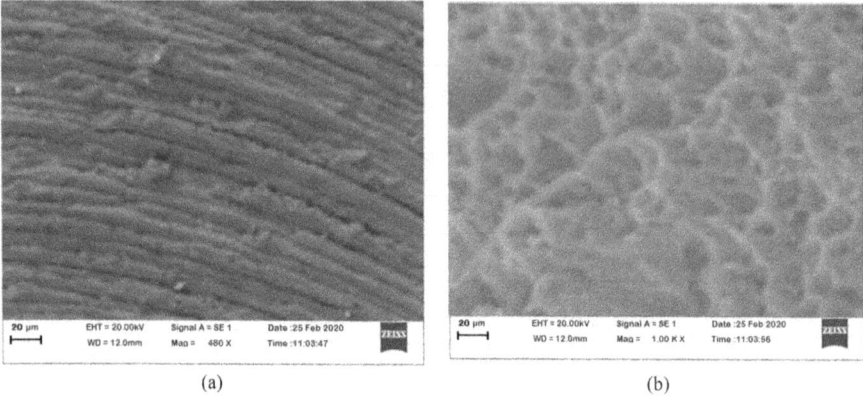

(a)

(b)

FIGURE 8.7   SEM image of stir welding zone.

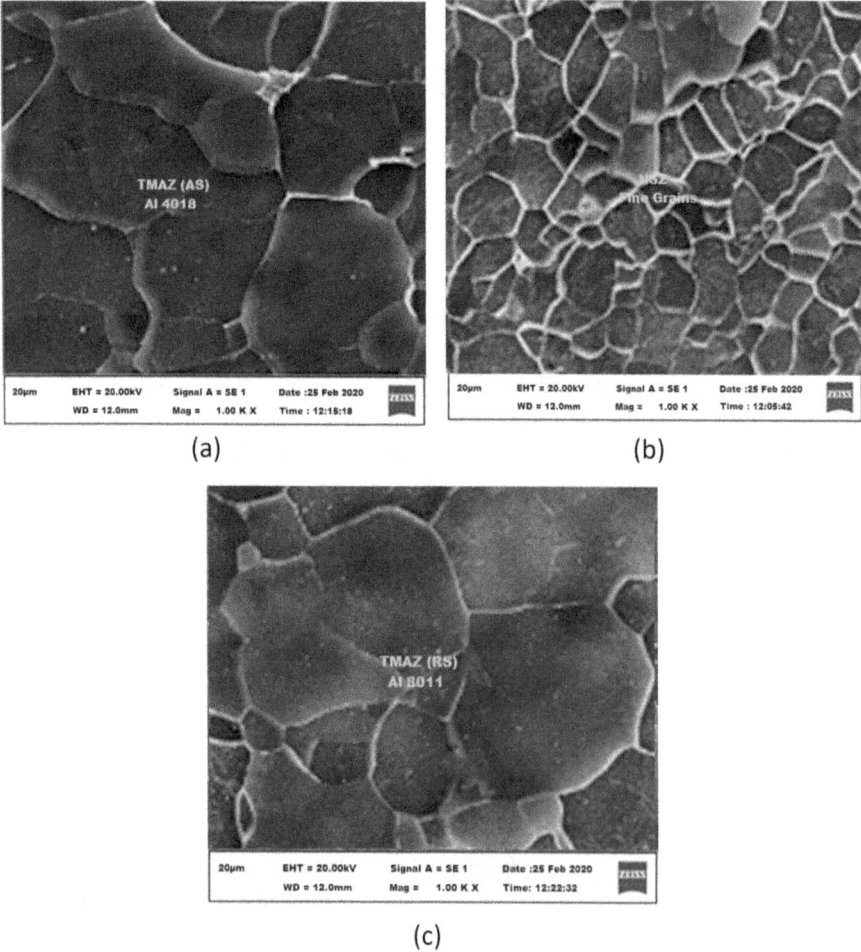

(a)

(b)

(c)

FIGURE 8.8   SEM image (a) TMAZ (AS), (b) NSZ, (c) TMAZ (RS).

8.6   OPTIMIZATION OF FUZZY TOPSIS METHOD

The TOPSIS method is technique to optimize a multi-objective problem. It is using parameters for order of preference by similarity to the ideal solution, but this technique is more efficient with a combination of Fuzzy techniques. Fuzzy method is a type of thinking prediction method (Sivapirakasam, Mathew, and Surianarayanan, 2011) and for forecasting an actual situation. Triangular and trapezoidal fuzzy numbers are commonly employed to model uncertainty in a variety of engineering situations. Although the study of triangular fuzzy numbers has been extensive, trapezoidal numbers are sporadically chosen because of their ease of computation. The advantage is that, compared to triangular fuzzy numbers Vs trapezoidal fuzzy numbers, the trapezoidal fuzzy is easy and mostly used for engineering application. This method can be used to simulate linear uncertainty in a verity of engineering applications.

A trapezoidal fuzzy number is used to designate the fuzzy linguistic variable, as seen in Figure 8.9. According to this graph, the linguistic values that are represent in Table 8.6 are multiplied with fuzzy weight (Dewangan, Gangopadhyay, and Biswas, 2015a). The five decision-makers provide their conclusions regarding the answers for each linguistic term characteristic weight that are given in Table 8.7.

Table 8.8 displays the respected fuzzy weight value of the decision-makers for each response. Table 8.9 displays the experimental value coupled with the

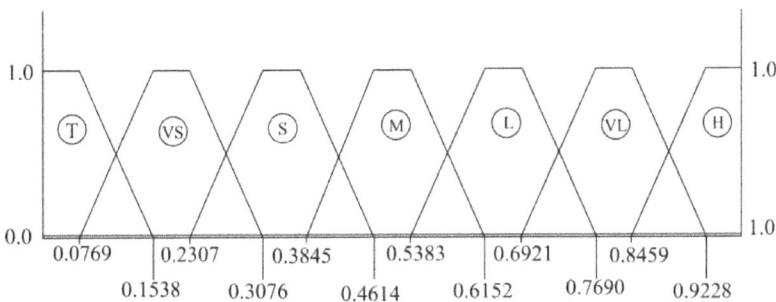

FIGURE 8.9   Trapezoidal membership function.

TABLE 8.6
Fuzzy Subset with Their Respected Fuzzy Weight

Fuzzy Subset	Respected Fuzzy Weight
Tiny (T)	(0.00, 0.00, 0.077, 0.154)
Very Small (VS)	(0.078, 0.154, 0.231, 0.308)
Small (S)	(0.231,0.308, 0.385, 0.462)
Medium (M)	(0.385, 0.462, 0.539, 0.615)
Large (L)	(0.539, 0.615, 0.692, 0.769)
Very Large (VL)	(0.692, 0.769, 0.846, 0.923)
Huge (H)	(0.846, 0.923, 1.00, 1.00)

TABLE 8.7

Decision Maker for Responses

Responses	Choice of Decision				
	DM-1	DM-2	DM-3	DM-4	DM-5
Ts	L	VL	H	VL	L
Ct	VL	H	H	L	L

TABLE 8.8

Fuzzy Weight

Responses	Fuzzy weight
Ts	0.6613, 0.7382, 0.8152, 0.8767
Ct	0.6921, 0.7690, 0.8460, 0.8922

TABLE 8.9

Normalized Matrix

Run No.	Experimental Calculated Value		R_{ij}	
	Ts (N)	Ct (N)	Ts	Ct
1	67	21	0.31600	0.09904
2	88	42	0.41504	0.19809
3	64	27	0.30185	0.12734
4	86	28	0.40561	0.13206
5	68	53	0.32072	0.24997
6	53	34	0.24997	0.16036
7	68	27	0.32072	0.12734
8	62	42	0.29242	0.19809
9	73	22	0.34430	0.10376

normalized response (R_{ij}). This equation can be used to calculate the normalized matrix, where R_{ij} is the normalized value.

$$R_{ij} = \frac{x_{ij}}{\sqrt{\sum_{i=1}^{9} x_{ij}^2}} \tag{8.1}$$

In this case, x_{ij} denotes the experimental value of the jth experimental run's i^{th} characteristic.

The remainder of the process involves multiplying all of the normalized matrix's (R_{ij}'s) characteristics by fuzzy weights. The weighted performance matrix, which is indicated by S_{ij}, is the resulting matrix (for i^{th} experimental run and j^{th} response), A similar approach was used by several researchers (Dewangan, Gangopadhyay, and Biswas 2015b; Kumar, Dewangan, and Pandey, 2020). Now, the following equations represent the positive ideal solution (S+) and the negative ideal solution (Equations 8.2 and 8.3).

$$S^+ = \left[\max \left(S_{ij} \right) \right] j \in J \text{ or } \left[\min \left(S_{ij} \right) j \in J \right], j = 1, 2, \ldots 9 \tag{8.2}$$

$i = 1$ and 2

$$S^- = \left[\min \left(S_{ij} \right) \right] j \in J \text{ or } \left[\max \left(S_{ij} \right) j \in J' \right] \tag{8.3}$$

J' is associated to the lower-the-better performances qualities, whereas J is related to the higher-the-better performances features. In this instance, J is taken into account because all of the responses (Ts and Ct) fall within the higher-the-better quality characteristics type. The normalized positive trapezoidal fuzzy numbers, with a range of 0 to 1, are the characteristics of S_{ij} (Table 8.10).

The next step to finding the distance, this distance is the difference between positive and negative ideal value and this calculation with the help of Equations 8.4 and 8.5 after calculating the S_{ij} matrix.

$$d_i^+ = \sum_{i=1}^{5} d\left(S_{ij}, S_j^+ \right), i = 1, 2 \ldots .9 \tag{8.4}$$

$$d_j^- = \sum_{i=1}^{5} d\left(S_{ij}, S_j^- \right), i = 1, 2 \ldots .9 \tag{8.5}$$

$$d(x, y) = \sqrt{\frac{1}{4} \left[\left(x_1 - y_1 \right)^2 + \left(x_2 - y_2 \right)^2 \right]} \tag{8.6}$$

TABLE 8.10
S_j^+ and S_j^- Values

		S_j^+ and S_j^-			
Ts	S_1^+	0.2745	0.3064	0.3383	0.3639
	S_1^-	0.1653	0.1845	0.2038	0.2191
Ct	S_2^+	0.1730	0.1922	0.2115	0.2230
	S_2^-	0.0685	0.0762	0.0838	0.0884

TABLE 8.11

(CCI) Closeness Coefficient Value

Run No.	di−	di+	CCI
1	0.3681	0.0954	0.2058
2	0.0774	0.3862	0.8331
3	0.3464	0.1171	0.2527
4	0.1894	0.2741	0.5913
5	0.1363	0.3272	0.7060
6	0.3721	0.0914	0.1972
7	0.3191	0.1444	0.3115
8	0.2545	0.2090	0.4509
9	0.3202	0.1433	0.3092

Where $d(x,y)$ is the separation of two ambiguous numbers. Equation 8.6 is used to determine the fuzzy number representing the distance between two trapezoidal numbers, and Table 8.11 displays the values for d+ and d−. The closeness coefficient (CCi), which is determined using Equation 8.7, is the last step after measuring the distance of ideal solution. CCi number represents how closely each experimental result comes to the ideal answer represented in Table 8.11.

$$CC_i = \frac{d_i^-}{d_i^+ + d_i^-} \tag{8.7}$$

Fuzzy-TOPSIS optimize the process parameters is second run experiment that is temperature 200°C, holding time 50 minutes and speed of tool rotation is 800rpm. The higher the close coefficient index (CCI) gives the optimum process parameters. Based on this close coefficient index value, significant parameters have been observed using ANOVA analysis. The significant parameters tested under 5% significance level is holding temperature have its P-value is 0.020 which is less than 0.05. The optimal set of technological parameters includes is temperature 250°C, holding time 50 minutes, and speed of tool rotation is 800 rpm as shown in Figure 8.10.

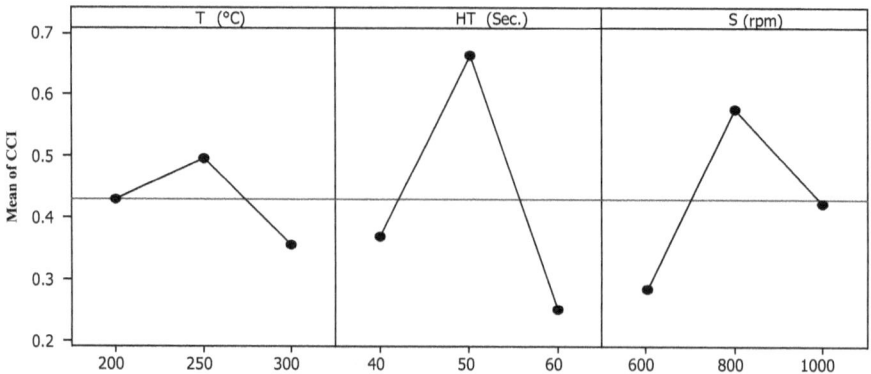

FIGURE 8.10   Optimum Parameter for Closeness coefficient.

While compare CCI optimize parameters results of the quality parameters at optimal conditions for best tensile strength and Charpy test.

8.7 CONCLUSION

Friction stir welding will be used to join the aluminum alloy AA1100 while using various process parameters to create a Taguchi design. We will examine their impact on the created joints' mechanical qualities in terms of tensile strength and the Charpy test.

- Welding joint of two similar grades has to be made successfully.
- The cooling medium has a significant impact on the tensile strength of the joints. Due to the increased heat input per unit length of the joint as welding speed rose, the joint's tensile characteristics degraded.
- Cooling of specimen is also affecting the material strength so it may be extended as future work.
- Optimum parameters are also verified with microstructure graph.
- Taguchi-based optimization method is the right solution for friction stir welding to the Fuzzy TOPSIS multi objective optimization techniques optimum parameter setting is temperature 250°C, holding time 50 minutes, and speed of tool rotation is 800rpm.

REFERENCES

Azadi, Chegeni Ava, and Platon Kapranos. 2018. "A Microstructural Evaluation of Friction Stir Welded 7075 Aluminum Rolled Plate Heat Treated to the Semi-Solid State." *Metals MDPI* 8 (1). https://doi.org/10.3390/met8010041

Burek, Rafał, Dawid Wydrzyński, Jarosław Sęp, and Wojciech Więckowski. 2017. "The Effect of Tool Wear on the Quality of Lap Joints between 7075 T6 Aluminum Alloy Sheet Metal Created with the FSW Method." *Eksploatacja I Niezawodnosc-Maintenance and Reliability* 20: 100–106.

Chen, Xinyi, Shujin Chen, Zhihong Liu, Yang Gao, and Hao Zhang. 2023. "Magnetic Field-Assisted Ultra-High-Speed Friction Stir Welding." *Materials Letters* 331: 133447. https://doi.org/10.1016/j.matlet.2022.133447

Dewangan, S., S. Gangopadhyay, and C. K. Biswas. 2015a. "Study of Surface Integrity and Dimensional Accuracy in EDM Using Fuzzy TOPSIS and Sensitivity Analysis." *Measurement: Journal of the International Measurement Confederation* 63: 364–76. https://doi.org/10.1016/j.measurement.2014.11.025

Dewangan, Shailesh, Soumya Gangopadhyay, and Chandan Kumar Biswas. 2015b. "Multi-Response Optimization of Surface Integrity Characteristics of EDM Process Using Grey-Fuzzy Logic-Based Hybrid Approach." *Engineering Science and Technology, an International Journal* 18 (3): 361–68. https://doi.org/10.1016/j.jestch.2015.01.009

Kumar, Prakash, Shailesh Dewangan, and Chandan Pandey. 2020. "Analysis of Surface Integrity and Dimensional Accuracy in EDM of P91 Steels." *Materials Today: Proceedings* 33: 5378–83. https://doi.org/10.1016/j.matpr.2020.03.119

Raj, Sanjay, and Pankaj Biswas. 2023. "Effect of Induction Preheating on Microstructure and Mechanical Properties of Friction Stir Welded Dissimilar Material Joints of Inconel 718 and SS316L." *CIRP Journal of Manufacturing Science and Technology* 41: 160–79. https://doi.org/10.1016/j.cirpj.2022.12.014

Rajeesh, J., R. Balamurugan, and K. Balachandar. 2018. "Process Parameter Optimization for Friction Stir Welding of Aluminium 2014-T651 Alloy Using Taguchi Technique." *Journal of Engineering Science and Technology* 13 (2): 515–23.

Sathish, S., J. Venkatesh, Pradeep Johnson, Shaik Annar, J. Saranya, J. Sai Chandra, Ramaswamy Subbiah, and S. Tharmalingam. 2022. "Strain Hardening Behaviour of Friction Stir Welded Magnesium Alloy." *Materials Today: Proceedings.* https://doi.org/10.1016/j.matpr.2022.10.217

Sivapirakasam, S. P., Jose Mathew, and M. Surianarayanan. 2011. "Multi-Attribute Decision Making for Green Electrical Discharge Machining." *Expert Systems with Applications* 38 (7): 8370–74. https://doi.org/10.1016/j.eswa.2011.01.026

Zhang, Z., Z. J. Tan, Y. F. Wang, D. X. Ren, and J. Y. Li. 2023. "The Relationship between Microstructures and Mechanical Properties in Friction Stir Lap Welding of Titanium Alloy." *Materials Chemistry and Physics* 296: 127251. https://doi.org/10.1016/j.matchemphys.2022.127251

9 Multi-Response Optimization of Process Parameters in Friction Stir Additive Manufacturing of Magnesium Alloy

Prakash Kumar Sahu
National Institute of Technology Agartala, Tripura, India

Chinmaya Prasad Nanda
KIIT Deemed University, Bhubaneswar, India

Jayashree Das
Indian Institute of Technology, Guwahati, India

Qingyu Shi
Tsinghua University, Beijing, China

CONTENTS

DOI: 10.1201/9781003346623-9

9.1   INTRODUCTION

Basically, the industrial revolution can be divided into four different stages. The first industrial revolution saw the utilization of steam power to increase productivity, which was minimal before due to the complete dependence of manufacturing on human labor. The second industrial revolution saw the use of electricity for mass production and a decrease in average manufacturing time. The third industrial revolution introduced the concept of automation in the manufacturing process by making the use of computer technology. The fourth industrial revolution, which we are currently in, has taken the idea of automation to a whole new level. The present research work is in this stage of industrial revolution so that the full potential of a machine capability is explored to make them interact with each other so there is coordination among the different machine parts that are to be used for the manufacturing of a certain component. These include the advent of Artificial Intelligence (AI), making the computer smarter and helping it learn complicated jobs and recognize complex patterns, faster computer processing, and Robotics, making the manufacturing process fully automated and one of the most important invention of the fourth industrial revolution, i.e., Additive Manufacturing Technology (AM). Also, Friction Stir Additive Manufacturing (FSAM) is one of the advanced processes which are used to develop novel alloys or composites for recent use as needed.

Now a days, Magnesium has a wide range of applications in the engineering field due to its low weight structure and higher strength. For this special reason, it is suitable to use in the fields of aerospace, automobiles, electronics, and biomedical industries to give high strength to the output product. There are various types of alloys like Aluminum, Zircon, Thorium, and rare earth metals used to give high strength to weight ratio. Basically, Magnesium alloys are used for both casting and forging processes to produce components. However, melting the Mg alloy is difficult because of its high oxidation and spattering. So FSAM is a good solid-state candidate since it worked at the plasticized state without melting the Mg and get different benefits. Zirconium based alloy is more advantageous because it can be used at a higher temperature, so aerospace components are made with Zirconium based alloy. One of the major advantages to mixing alloying material with magnesium (like manganese) is to improve corrosion resistance. Magnesium alloy is 33% lighter than Aluminium and 75% lighter than steel, and it provides high strength with high rigidity to the material. Magnesium alloys have two major disadvantages, i.e., a low high temperature (low creep strength) and another low corrosion resistance. So, to avoid this a specific amount of Magnesium mix to Magnesium matrix is used to prepare a Magnesium alloy. If the amount of Magnesium is more than 50% in the mixture, then it is more brittle and easily corrosive, but an amount of 5% external alloy gives more strength, corrosion resistance, hardness, etc. The basic purpose of the process is to change the heterogeneous microstructure to homogeneous microstructure or modified grain structure to achieve better mechanical and microstructural properties. This process overcomes various problems which are generated during traditional joining processes. The adopted FSAM is a type of Friction stir processing technique. By this method, the intense properties of the metal are changed, and a plastic deformation takes place to complete the output product.

Friction stir processing is a next generation process of friction stir welding (FSW), where the material top surface is processed instead of welding. It is a method where the material properties are changed through intense localized plastic deformation. In this process, a tool is inserted into the surface of the workpiece and the tool is rotating with high speed along the direction of the processing. With this processing, composite material is to be produced by adding reinforcement particles in the form of a grove below the tool to improve the efficiency, hardness, and corrosion resistance of the material. The FSP process was invented by Wayne Thomas after the year 1991, and it helps the manufacturer to make the component with high accuracy for many industries to a value-added product. The FSAM technology was first introduced around 2006 when Boeing and Airbus demonstrated how structures can be fabricated at a faster rate with minimal material waste. In Friction Stir Additive Manufacturing a non-consumable rotating tool is fitted with a custom designed shoulder and pin which is meant to be inserted into the surface of sheets and plates to be joined along the line of traverse. Similarly, many layers of workpiece are joined layer by layer to construct a component of a desired shape and size. The joints that are produced by this process are in solid state and do not involve melting at any stage, thus eliminating the emergence of defects that are usually caused by melting, such as cracking, porosity, anisotropy, and residual stresses due to the temperature of the process. Also, the FSAM process is able to produce parts that have finer grain microstructure as opposed to the other fusion based techniques.

In this article it has been observed much research work concerns magnesium alloy metal with friction stir processing and friction stir additive manufacturing for its effect on tool rotational speed, traverse speed, tilt angle, shoulder diameter, pin diameter and plunge depth, etc. From the various paper studies, it was found that Jamili et al. (2021) investigated a rare earth bearing magnesium alloy processed by FSP and cold compression (i.e., compressive load down to 10% of engineering strain) over the grain structure for grain refinement, distribution and micro texture of the composite material. The purpose is to modify and improve the microstructure and texture of the functional graded material. Hassanamraji et al. (2021) used a thermo-mechanical finite element method to develop AZ91 material find the temperature and strain rate for conical and cylindrical pin. Also, he found more displacement of materials occurred in case of conical pin compare with cylindrical pin, the lateral force is managed with the flow of velocity and pressure in FSP. Harwani et al. (2021) studied that during dynamic recrystallization, the material properties and grain size is changed and to get the maximum ductility in plastic deformation with high temperature leads to super-plasticity due to the elongation during the process by FSP. Nasiri et al. (2021) observed the grain refinement and that there will be changes in the intermetallic phases with uniform distributed fined particles to enhance the mechanical properties of the material processed by FSAM. Srivastava et al. (2021a, 2021b) studied with the help of FSP that the lightweight aluminum and magnesium alloy-based material get the better grain refinement and good mechanical properties can be achieved by the material. The various elements for this process consist of a rotating tool, shoulder, pin, workpiece, clamping device, vertical CNC milling machine, etc.

With further review, it was found that tool rotational speed is the speed at which the tool rotates on its own axis when it is inserted inside of the workpiece and is measured in revolution/minute (RPM), thus, an FSP process involves a tool rotating at high speed. Parameters play an important role while producing magnesium-based material processed by FSP. Luo et al. (2018) made use of solid state FSP of AZ61 magnesium alloy with increasing rotational speed and traverse speed to increase micro hardness and tensile strength of the material. The result shows that the grain size of magnesium alloy material was obtained by FSP, and it is controlled by tool rotation rate and traverse speed. Vedabouriswarn et al. (2018) investigate choosing the parameters in FSP that the tool rotation speed and traverse speed with tool shoulder diameter and pin diameter generate the heat between the tool and the surface of the workpiece. The main purpose is to select specific parameters with a specific amount to produce a RZ5 magnesium alloy material with a developed metal matrix composite having good mechanical properties. Yousefpour et al. (2021) study the various effects on microstructure and texture due to the speed of the tool. For the experimental work, the shoulder diameter was 16 mm with a rotational speed of 400, 800, and 1200 rpm. The result was that there will be an increase of rotational speed from 400 to 1200 rpm, and the value of UTS increased 236.8–267.1 MPa. Sagar et al. (2021) is manufacturing AZ61Mg composite material using a triangular tool operated by FSP. Considering various fixed parameters like tool rotation speed 1250 rpm, tilt angle 3°, and number of passes 4 with changes of traverse speed 20, 40, and 63 mm/ min. The results show that whenever there is a repeat of the number of passes, hardness value increases, as well as the grain size is reduced to get the better product.

9.2   EXPERIMENTAL DETAILS

Magnesium plate is taken as the base plate for this present experiment based on FSAM. For this experiment, AZ31 magnesium alloy having the dimension of 100 × 100 × 4 mm plate is used to process the material at different passes of the tool, different Al% and other process parameters are considered. The chemical composition for AZ31 plate with AL 2.50 to 3.50%, Zn-0.60 to 1.40 % and Mn-0.20% and balanced with Mg is detected by the EDX (Energy Dispersive X-Ray Analysis) test. During operation there are many independent parameters which can be controlled by this experiment. For this experiment, the tool has the two primary functions, (i). Heat generation in between the shoulder and workpiece. (ii). Deformation of workpiece material through the rotating pin. To conduct the experiment, an H13 tool is used for the process. The rotating pin stirs the heated material around it and fills the cavity at the backside of the tool. The material flows due to severe plastic deformation and thermal heating; due to this, the material undergoes a significant grain refinement. The grain refinement occurs on the base material to improve material properties of the first material while mixing with the supplied second material. With the help of this process, the two different materials come to bond as alloy without heating, melting, or changing the material properties. A special fixture is to be made to hold the workpiece rigidly, maintaining proper position during operation. The fixture is designed in such a way that it can hold thin plates rigidly with proper position. The main objective is to make this clamping device clamp the workpiece metal in a proper position during the movement of the tool, as well as the worktable in X, Y, and Z directions. In this experiment, four major parameters are to be considered.

These are tool rotational speed, tool traverse speed, percentage of Aluminum (% of Al), and the number of passes. During the experiment, various independent parameters are to be controlled by this experiments.

Taguchi's L16 design method is used to minimize the number of experiments for this work. The input process parameters with values are given in Table 9.1. The present investigation considered four important process parameters: tool rotational speed, tool traverse speed, percentage of aluminum powder as additives, and number of passes of the tool. Four process parameters are considered in each four level so that the Taguchi's L16 design of matrix can be seen in Table 9.2. After getting the design matrix, we performed the experiments as given in the Table 9.2. With completion of the experiment, we have to move for the extraction of the different sample for the mechanical analysis. The tested different important outputs are ultimate tensile

TABLE 9.1
Process Parameters Value for Different Levels

Level of Process Parameters	Tool Rotational Speed (rev/min)	Tool Traverse Speed (mm/min)	% of Al	Number of Passes of Tool
1	450	30	0	1
2	600	45	3	2
3	750	60	6	3
4	900	75	9	4

TABLE 9.2
Taguchi L16 Design Matrix with Parametric Value

Experiment Number	Tool Rotational Speed (rev/min)	Tool Traverse Speed (mm/min)	% of Al	Number of Passes of Tool
1	450	30	0	1
2	450	45	3	2
3	450	60	6	3
4	450	75	9	4
5	600	30	3	3
6	600	45	0	4
7	600	60	9	1
8	600	75	6	2
9	750	30	6	4
10	750	45	9	3
11	750	60	0	2
12	750	75	3	1
13	900	30	9	2
14	900	45	6	1
15	900	60	3	4
16	900	75	0	3

TABLE 9.3

Output Value Calculated According to the Parameters Setting of Table 9.2

Exp. No	UTS (MPa)	YS (MPa)	Percentage of E (% E)	Max. H at NZ	Avg. H at NZ
1	120.8	77.09	3.10	65.57	55.98
2	184.6	135.6	6.44	91.87	74.65
3	190.7	140.8	7.42	94.65	75.81
4	180.2	110.6	3.52	83.42	65.41
5	157.9	118.7	4.45	80.18	68.19
6	187.9	127.9	3.84	88.08	72.31
7	205.6	130.5	4.84	97.63	73.22
8	210.4	172.8	7.92	106.71	82.53
9	229.5	170.7	7.91	98.82	78.69
10	249.4	175.5	10.04	111.24	87.11
11	193.8	142.6	7.04	95.84	72.86
12	174.4	125.5	5.04	84.17	70.96
13	186.5	121.8	3.54	89.93	65.78
14	200.8	126.9	3.93	96.54	72.90
15	170.4	132.8	4.08	81.95	70.66
16	167.6	110.7	3.76	78.85	63.79

strength (UTS), yield strength (YS), percentage of elongation (%E), maximum hardness at NZ and average hardness at NZ, which is given in Table 9.3.

9.3 TAGUCHI GREY RELATIONAL ANALYSIS PROCESS

Taguchi Grey Relational Analysis (GRA) is a method to determine the optimized process parameters setting to get the desired output. For the present work, we need the best process parameter that gives a fine structure to the material while enhancing the mechanical properties. Basically, this method is used to calculating the process parameters setting in a reduced way for the process without considering full factorial experiments. This design method used a mathematical equation to calculate the ratio between mean (signal) to the standard deviation (noise) called S/N ratio. Basically, the signal-to-noise ratio is effective in minimizing the uncontrolled factor to get the best process or product of the process. The S/N ratio is divided in three types for the calculation, these are **Nominal the best (NB)**, **Higher the best (HB)** and **Lower the best (LB)**. Previously, the Taguchi traditional method was used for optimization process with single output but Taguchi GRA method can also be used for a multi objective optimization problem. The GRA system is introduced to the manufacturing sector in the year 1982 by scientist Deng. In the GRA method, firstly, the experimental data go for the Normalization process in the range of zero to one; the nest step is normalized data calculate grey relational coefficient and find a relation between the desired experiment and actual experiment. After that, calculate the average value of grey relational coefficient, called grey relational grade (GRG).

9.4   RESULT AND DISCUSSION

With the help of the Taguchi GRA method, optimize the process parameter processed by friction stir additive manufacturing. This method can utilize multi-functional problems into a single functional problem. This technique can also find out various parameters with the best value for both cases, i.e., higher the better (HB) and lower the better (LB). ANOVA method is used for analysis of variance and to predict the optimal GRA.

9.4.1   CALCULATION WITH TAGUCHI GRAY RELATIONAL ANALYSIS (GRA)

After given the experimental data (Table 9.1), one matrix table is to be formed (Table 9.2) with the help of the Taguchi L16 design method. Using GRA, the first step is to normalize the experimental output data as given in Table 9.3. Then, calculate the Grey relational generation. After the calculation of Grey relational generation, optimize the Grey relational coefficient (GRC). Finally the Grey relational grade (GRG) is optimized to taking the average data of GRC. After the experimental testing, the output values are represented in Table 9.3. Then, with continuation of this table we have to move forward for normalization to the output and to get the GRC and GRG, etc.

Different formulae are given below to find out the Lower is the better (**LB**) and Higher is the better (**HB**) value. Subsequently, a researcher can chose any one of the given conditions according to their requirement. The selection of **LB** and **HB** purely depend on the researcher's requirement or any past experience. For the calculation of normalization, experimental data is to be considered with the range from 0 to 1 by following the equation as given below. In this work, we also considered two different cases: **CASE-1** (Consider **LB**) and **CASE-2** (Consider **HB**). After that, a comparison study can be performed, with some conclusions.

CASE-1: Experimentation outputs are normalized between 0 and 1. If we wish to minimize output then consider **LB**. To minimize the response, LB category is used for normalization, using Equation 9.1. Then the equation is

$$xi(k) = \frac{\max yi(k) - yi(k)}{\max yi(k) - \min yi(k)}.. \tag{9.1}$$

CASE-2: If we wish to maximize output response, then consider **HB**. For maximization, the response HB category is used for normalization, using Equation 9.2. Then the equation is

$$xi(k) = \frac{yi(k) - \min yi(k)}{\max yi(k) - \min yi(k)} \tag{9.2}$$

Where xi(k) = after the calculation GRG (Grey relational generation), min yi(k) is least value of yi(k) for Kth response. Max yi (k) is largest value of yi(k) for kth response. I = 1,2,3...... indicates experiments number and k = 1,2,3....... indicates responses number.

After the calculation of **HB** and **LB**, then calculate **GRC** (grey relational coefficient). GRC can be calculated to recognize the correlation among the reference order and compatibility order. GRC can be calculate with the use of Equation 9.3.

$$\xi = \frac{\Delta_{min} + \psi \Delta_{max}}{\Delta_{oi}(k) + \psi \Delta_{max}}$$

(9.3)

Where $\Delta_{oi}(k)$ = absolute difference value of $x_o(k)$ and $x_i(k)$. ψ is the coefficient; $0 \le \psi \le 1$, Δ_{min} is the minimum value of $\Delta_{oi}(k)$ Δ and Δ_{max} is the maximum value of $\Delta_{oi}(k)$.

The **GRG** (grey relational grade) (γ) is the mean of GRC and it gives the relation between the sequence. It can be calculate using Equation 9.4. It gives evidence for the association between the sequences. Its calculated output will be in between 0 and 1.

$$\gamma_i = \frac{1}{n} \sum_{k=1}^{n} \xi_i(k)$$

(9.4)

Where n = is process response number. Optimum GRG found by using Equation 9.5.

$$\gamma_e = \gamma_m + \sum_{i=1}^{q} (\overline{\gamma}_i - \gamma_m)$$

(9.5)

Where γ_m is the total mean of GRG, q = number of input parameters, and $\overline{\gamma}_i$ is the mean GRG value at the optimal level for the ith parameter. After completion of GRG calculation ANOVA. Taguchi ANOVA method is used to optimize the most significant value for each parameter with percentage contribution of all process parameters.

9.4.2   GETTING OPTIMAL PARAMETERS WITH THE HELP OF TAGUCHI GRA

In this work, Taguchi GRA method is used to optimize the output to get the best parameters setting among these process parameters. As discussed earlier, Table 9.3 represents the experimental output found after the experiments. Taking these 5 output parameters, the normalization can be calculated with the help of either Equation 9.1 or Equation 9.2, depending on the researcher's requirements. Five output parameters, i.e, Ultimate tensile strength (UTS), yield strength (YS), % E, average hardness value at NZ, and Maximum hardness value at NZ are taken to get the normalization data. In this case, the researcher considers two cases, like **Case 1** represents all output can be considered as *"Lower the better"* and **Case 2** represents all output as *"Higher the better."*

CASE-1:   ALL THE MECHANICAL TESTING OUTPUT TAKEN AS "LOWER THE BETTER"

Considering **Case-1**, all the testing output, i.e., UTS, YS, % of E, Max H at NZ, and Avg. H at NZ are considered lower the better. Using Equation 9.1, calculate the normalized value for these output values given in Table 9.4. After the calculation of normalization, then GRC (grey relational coefficient) can be calculated, and the calculated value is given in Table 9.5. The GRC can be calculated by the Equation 9.3 and the GRGs can be calculated by using Equation 9.4. The GRG can convert the multi output to get the single output for the calculation of S/N ratio. The GRG value can be varied from 0 to 1 and this GRG can give the S/N ratio as represented in the Table 9.6.

TABLE 9.4
Normalization Value for "Lower the Better"

EXP. No.	UTS (MPa)	YS (MPa)	Percentage of E (% E)	Max. H at NZ	Avg. H at NZ
1	1	1	1	1	1
2	0.503888025	0.405446601	0.518731988	0.424129626	0.400256987
3	0.456454121	0.352606442	0.377521614	0.363258156	0.362993897
4	0.538102644	0.659485825	0.939481268	0.609152617	0.697076775
5	0.711508554	0.577177116	0.805475504	0.680096343	0.607773852
6	0.478227061	0.483690682	0.893371758	0.507116269	0.475425634
7	0.34059098	0.457270603	0.749279539	0.298007445	0.446193383
8	0.303265941	0.027436236	0.305475504	0.09918984	0.14712496
9	0.15474339	0.048775531	0.306916427	0.271950952	0.270478638
10	0	0	0	0	0
11	0.432348367	0.334315618	0.432276657	0.337201664	0.45775779
12	0.583203733	0.508078447	0.720461095	0.592730458	0.518792162
13	0.48911353	0.545676252	0.936599424	0.466608277	0.685191134
14	0.377916019	0.493852251	0.880403458	0.321874316	0.456472856
15	0.614307932	0.433898994	0.858789625	0.641340048	0.528429168
16	0.636080871	0.658469668	0.904899135	0.709218305	0.749116608

TABLE 9.5
GRC Calculation according to Normalization Value of Table 9.4

EXP. No.	UTS (MPa)	YS (MPa)	Percentage of E (% E)	Max. H at NZ	Avg. H at NZ
1	1	1	1	1	1
2	0.5019516	0.456807316	0.509544787	0.464740002	0.454651672
3	0.479135618	0.43577027	0.445442875	0.439853607	0.439751377
4	0.519805982	0.594873965	0.892030848	0.561263365	0.622724545
5	0.634122288	0.541815779	0.719917012	0.609827747	0.56039604
6	0.48934551	0.491976204	0.824228029	0.503583637	0.488007525

(Continued)

TABLE 9.5   (Continued)
GRC Calculation according to Normalization Value of Table 9.4

EXP. No.	UTS (MPa)	YS (MPa)	Percentage of E (% E)	Max. H at NZ	Avg. H at NZ
7	0.431254192	0.479510793	0.666026871	0.415975954	0.474470355
8	0.417803769	0.339543871	0.418576598	0.356936303	0.369583284
9	0.371676301	0.344536638	0.419082126	0.407149862	0.406662312
10	0.333333333	0.333333333	0.333333333	0.333333333	0.333333333
11	0.468317553	0.428932572	0.4682861	0.429997175	0.479734936
12	0.545377439	0.50407212	0.641404806	0.551104139	0.509576035
13	0.494615385	0.523931214	0.887468031	0.483843628	0.613640844
14	0.445599446	0.496944907	0.806976744	0.424402937	0.47914422
15	0.56453029	0.468998713	0.779775281	0.58230269	0.514630517
16	0.578757876	0.594155648	0.840193705	0.632285754	0.665882353

TABLE 9.6
GRG Value and Its S/N Ratio

EXP. No.	GRG	S/N Ratio(dB)
1	1	0.00000
2	0.477539075	−6.41982
3	0.44799075	−6.97462
4	0.638139741	−3.90168
5	0.613215773	−4.24773
6	0.559428181	−5.04511
7	0.493447633	−6.13518
8	0.380488765	−8.39316
9	0.389821448	−8.18269
10	0.333333333	−9.54243
11	0.455053667	−6.83875
12	0.550306908	−5.18790
13	0.60069982	−4.42685
14	0.530613651	−5.50443
15	0.582047498	−4.70083
16	0.662255067	−3.57949
	Average GRG is 0.544648832	Average S/N ratio is −5.567541875

The ANOVA analysis can be represented in the Table 9.7. This table represents probability of significance of each parameters as well as contribution to the entire experiments. The ANOVA also gives the total variances into the contribution of the whole parameter and their residual error. Considering Table 9.7 value, it was found that RPM (Tool Rotational Speed) has the maximum contribution of 32.24% in total value. Percentage of AL (% of AL) has the contribution of 31.83% and traverse speed (TS), number of pass (NP) has the lowest contribution in total. Also, find out there will be less error contribution compared with RPM contribution. The mean of each parameter at different level also calculated and represented in the Table 9.8.

TABLE 9.7
ANOVA for S/N Ratio from the GRG Data

Source	DF	Sum of Squares	Variance	Mean Sum	F-Value	Probability of Significance	Percentage of Contribution
RPM	3	24.898	24.898	8.2994	12.63	0.033	32.24
TS	3	13.603	13.603	4.5342	6.90	0.074	17.61
% OF Al	3	24.579	24.579	8.1930	12.46	0.034	31.83
NUMBER OF PASS	3	12.155	12.155	4.0518	6.16	0.085	15.74
RESIDUAL ERROR	3	1.972	1.972	0.6574			2.5
TOTAL	15	77.207					

TABLE 9.8
S/N Ratio (LB) Response Table

Level	RPM	TS	% of AL	Number of Pass
1	−4.324*	−4.214*	−3.866*	−4.207*
2	−5.955	−6.628	−5.139	−6.520
3	−7.438	−6.162	−7.264	−6.086
4	−4.553	−5.266	−6.002	−5.458
Delta	3.114	2.414	3.398	2.313
Rank	4	3	1	2

* Optimal level of parameters (RPM 1, TS 1, %Al 1, NP 1)

TABLE 9.9
Result for Final Experiment to Get LB

Output	Initial Parameter Setting of Exp 1 (RPM1, TS1, % of Al1, NP1)	Prediction (RPM 1, TS 1, %Al 1, NP 1)	Confirmation Experiment (RPM 1, TS 1, %Al 1, NP 1)
UTS (MPa)	120.8		121
YS (MPa)	77.09		77.1
Percentage of E	3.10		3.12
MAXIMUM HARDNESS AT NZ	65.57		65.6
AVERAGE HARDNESS AT NZ	55.98		56
S/N RATIO	0.00000	−1.11729	−1.11729

The result for the final experiment after the optimized condition is represented in Table 9.9. It was observed that the initial experiment 1 with considering **lower is better** is the same as the predicted experiment as represented in Table 9.9. So, there

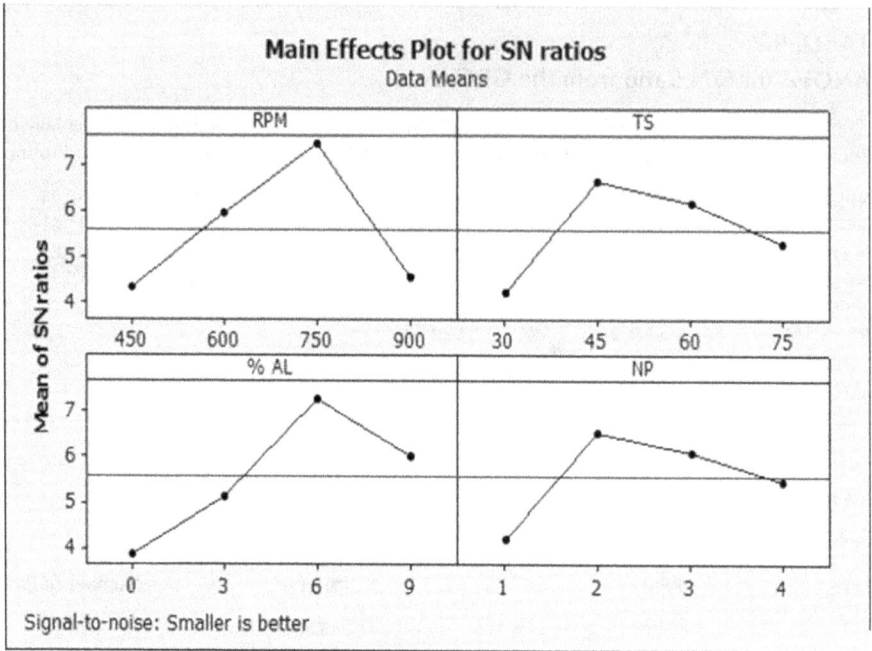

FIGURE 9.1   Signal to Noise ratio graph.

is no such variation with the optimized value and the calculated value. The mean effect plot is also represented in Figure 9.1.

CASE-2:   ALL THE MECHANICAL TESTING OUTPUT TAKEN AS "HIGHER THE BETTER"

Considering **Case-2**, all the testing output, i.e., UTS, YS, % of E, Max. H at NZ, and Avg. H at NZ are considered as higher the better. Using Equation 9.2, calculate the normalized value for these parameters given in Table 9.10. After the calculation of normalization, then GRC value is given in Table 9.11. For the calculation of S/N ratio, consider all the output values in the case of **Higher is better**. The values are given in **Table 9.12**. The GRG values varied in between 0 and 1. Figure 9.2 represent Signal to Noise ratio graph with considering higher is better.

TABLE 9.10
Normalization Value for Higher the Better

EXP. No.	UTS (MPa)	YS (MPa)	Percentage of E (% E)	Max. H at NZ	Avg. H at NZ
1	0	0	0	0	0
2	0.496111975	0.594553399	0.481268012	0.575870374	0.599743013
3	0.543545879	0.647393558	0.622478386	0.636741844	0.637006103
4	0.461897356	0.340514175	0.060518732	0.390847383	0.302923225
5	0.288491446	0.422822884	0.194524496	0.319903657	0.392226148
6	0.521772939	0.516309318	0.106628242	0.492883731	0.524574366
7	0.65940902	0.542729397	0.250720461	0.701992555	0.553806617
8	0.696734059	0.972563764	0.694524496	0.90081016	0.85287504
9	0.84525661	0.951224469	0.693083573	0.728049048	0.729521362
10	1	1	1	1	1
11	0.567651633	0.665684382	0.567723343	0.662798336	0.54224221
12	0.416796267	0.491921553	0.279538905	0.407269542	0.481207838
13	0.51088647	0.454323748	0.063400576	0.533391723	0.314808866
14	0.622083981	0.506147749	0.119596542	0.678125684	0.543527144
15	0.385692068	0.566101006	0.141210375	0.358659952	0.471570832
16	0.363919129	0.341530332	0.095100865	0.290781695	0.250883392

TABLE 9.11
GRC Value Calculate according to Normalization Table 9.9

EXP. No.	UTS (MPa)	YS (MPa)	Percentage of E (% E)	Max. H at NZ	Avg. H at NZ
1	0.333333333	0.333333333	0.333333333	0.333333333	0.333333333
2	0.498063517	0.55221368	0.490806223	0.541049639	0.555396967
3	0.522764228	0.586437042	0.569786535	0.579201015	0.579378373
4	0.48164794	0.431225626	0.347347347	0.450794591	0.417684154
5	0.412708601	0.464176218	0.383002208	0.42369422	0.451355662
6	0.511128776	0.508289861	0.358841779	0.496467007	0.51259674
7	0.594819611	0.522318348	0.400230681	0.626560571	0.528433203
8	0.622458858	0.94798189	0.620751342	0.834460077	0.7726483
9	0.763657957	0.911119341	0.619642857	0.647709545	0.648947259
10	1	1	1	1	1
11	0.536280234	0.599293587	0.536321484	0.597227671	0.522052658
12	0.461593683	0.495993146	0.409681228	0.457569382	0.490777235
13	0.505503145	0.478159467	0.348044132	0.517272624	0.421872883
14	0.569530558	0.503092889	0.362212944	0.608365526	0.522753988
15	0.448709002	0.535389805	0.367974549	0.438081535	0.486178354
16	0.440109514	0.431603877	0.355897436	0.413490267	0.400282885

TABLE 9.12
GRG Value and Its S/N Ratio

EXP. No.	GRG	S/N Ratio
1	0.333333333	−9.54243
2	0.527506005	−5.72251
3	0.567513439	−4.94934
4	0.425739932	−7.84982
5	0.426987382	−7.55885
6	0.477464833	−6.93691
7	0.534472483	−6.21016
8	0.759660093	−2.89061
9	0.718215392	−2.94306
10	1	0.00000
11	0.558235127	−5.25529
12	0.463122935	−6.86759
13	0.45417045	−7.52830
14	0.513191181	−6.67038
15	0.455266649	−7.10640
16	0.408276796	−8.02948
	Average GRG is 0.520679715	Average S/N ratio is −6.003820625

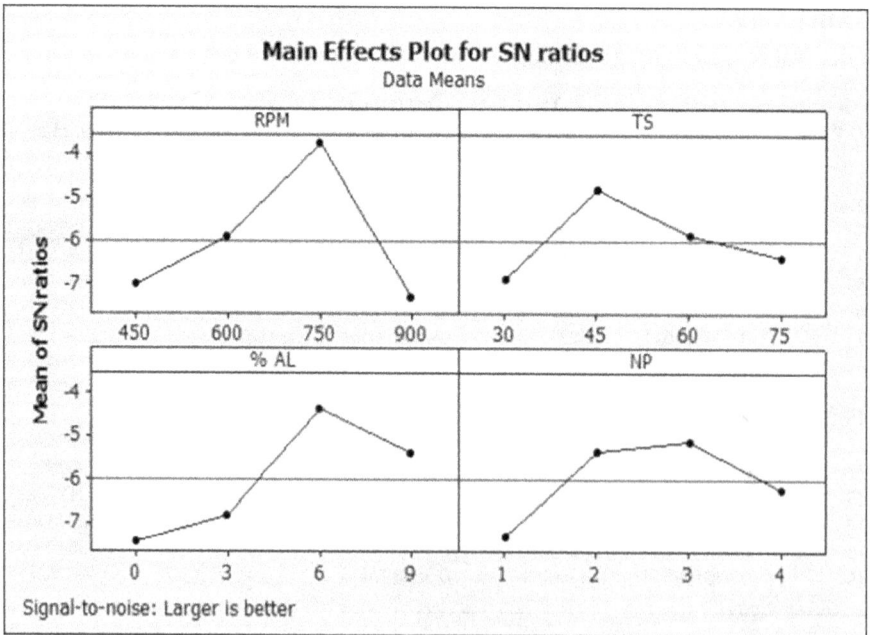

FIGURE 9.2   Signal to Noise ratio graph.

TABLE 9.13
ANOVA Analysis for S/N Ratio

Source	DF	Sum of Squares	Variance	Mean Sum	F-Value	Probability of Significance	Percentage of Contribution
RPM	3	31.238	31.238	10.413	3.34	0.174	36.77
TS	3	9.371	9.371	3.124	1.00	0.499	11.03
% OF AL	3	23.124	23.124	7.708	2.47	0.238	27.22
NUMBER OF PASS	3	11.863	11.863	3.954	1.27	0.425	13.96
RESIDUAL ERROR	3	9.355	9.355	3.118			11.01
TOTAL	15	84.952					

TABLE 9.14
S/N Ratio (HB) Response Table

Level	RPM	TS	% of AL	Number of Pass
1	−7.016	−6.893	−7.441	−7.323
2	−5.899	−4.832*	−6.814	−5.349
3	−3.766*	−5.880	−4.363*	−5.134*
4	−7.334	−6.409	−5.397	−6.209
Delta	3.567	2.061	3.078	2.188
Rank	1	3	2	4

* Optimal level of parameters (RPM3, TS2, % of Al 3, NP3)

Considering Table 9.13 value, it was found that RPM (Tool Rotational Speed) has the maximum contribution in total value. Percentage of AL (% of AL) has the contribution of 27.22% and traverse speed (TS), number of pass (NP) has the lowest contribution in total. Also find out there will be less error contribution compared with RPM contribution. In Table 9.14, mean S/N ratio for every parameter at various levels is shown. From the above two case consideration, in first case, i.e., case-1 (Lower the Better) has the much smaller error compared with case-2 (Higher the Better). However, the prediction for Case 2 is perfect with respect to Case 1.

Form the Table 9.15, it was found that at the initial parameter, i.e., revolution per minute in level three, traverse speed in level two, the percentage of aluminum in level four and number of passes in level three get a maximum UTS is 249.4 MPa. But a confirmation experiment gives a maximum UTS is 250.1 MPa and it's quite similar. So, the confirmation experiments give RPM in level three, TS in level two, % of AL in level three, and the number of pass in level three.

TABLE 9.15
Result for Final Experiment to Get Higher is Better

Output	Initial Parameter Setting of Exp 10 (RPM3, TS2, % of Al4, NP3)	PREDICTION (RPM3, TS2, % of Al 3, NP3)	CONFIRMATION EXPERIMENT (RPM3, TS2, % of Al 3, NP3)
UTS (MPa)	249.4		250.1
YS (MPa)	175.5		176.2
Percentage of E (% E)	10.04		10.05
MAXIMUM HARDNESS AT NZ	111.24		111.25
AVERAGE HARDNESS AT NZ	87.11		87.15
S/N RATIO	0.00000	−1.11896	−1.11896

9.5 CONCLUSION

Current work is discussed about the process parameter optimization with the help of Taguchi GRA and ANOVA design for the AZ31 Mg alloy material processed by Friction stir additive manufacturing processing. In this paper, we took different percentages of Al (0, 3, 6, 9%) with Magnesium base plate to make the composite material. With the multi-pass of the tool there will be an improvement in structural and mechanical properties for the material. For the experiment, various rotational speeds 450, 600, 750, and 900 rpm and traverse speed 30, 45, 60, and 75 mm/min has been considered for the process. Also, various percentages of Aluminum additives like 0, 3, 6, and 9 % is to be added with Magnesium alloy and processed by FSAM where the tool pass is to be considered.

From the paper, various conclusions are found. These are as follows:

- With the help of ANOVA, it was found that tool rotational speed is the most important parameter, while choosing lower is the better as well as higher is the better with the highest percentage of contribution, and the percentage of error is also less. It was found that the Taguchi grey relational method is the best method to optimize the best parameter of multi optimization problem in the FSAM process.
- Friction Stir Process proves to be a significant improvement as compared to the conventional manufacturing processes as it improves the mechanical properties and makes the component more durable to stresses.
- The process parameters play an important role in the final properties of the manufactured component by FSAM.
- Choosing the higher is better and the result shows that with Taguchi GRA there will be an improvement in all the mechanical properties considered in this work.

REFERENCES

Harwani, D., Badheka, V., Patel, V., Li, W., & Andersson, J. (2021). Developing superplasticity in magnesium alloys with the help of friction stir processing and its variants–A review. *Journal of Materials Research and Technology*, *12*, 2055–2075.

Hassanamraji, N., Eivani, A. R., & Aboutalebi, M. R. (2021). Finite element simulation of deformation and heat transfer during friction stir processing of as-cast AZ91 magnesium alloy. *Journal of Materials Research and Technology*, *14*, 2998–3017.

Jamili, A. M., Zarei-Hanzaki, A., Abedi, H. R., & Minarik, P. (2021). Development of grain size/texture graded microstructures through friction stir processing and subsequent cold compression of a rare earth bearing magnesium alloy. *Materials Science and Engineering: A*, *814*, 141190.

Luo, X. C., Zhang, D. T., Zhang, W. W., Qiu, C., & Chen, D. L. (2018). Tensile properties of AZ61 magnesium alloy produced by multi-pass friction stir processing: Effect of sample orientation. *Materials Science and Engineering: A*, 725, 398–405.

Nasiri, Z., Khorrami, M. S., Mirzadeh, H., & Emamy, M. (2021). Enhanced mechanical properties of as-cast Mg-Al-Ca magnesium alloys by friction stir processing. *Materials Letters*, *296*, 129880.

Sagar, P., & Handa, A. (2021). Selection of tool transverse speed considering trial run experimentations for AZ61/Tic composite developed via friction stir processing using triangular tool. *Materials Today: Proceedings*, *38*, 198–203.

Srivastava, A. K., Dixit, A. R., Maurya, M., Saxena, A., Maurya, N. K., Dwivedi, S. P., & Bajaj, R. (2021a). 20th Century uninterrupted growth in friction stir processing of lightweight composites and alloys. *Materials Chemistry and Physics*, *266*, 124572.

Srivastava, A. K., Saxena, A., & Dixit, A. R. (2021b). Investigation on the thermal behaviour of AZ31B/waste eggshell surface composites produced by friction stir processing. *Composites Communications*, *28*, 100912.

Vedabouriswaran, G., & Aravindan, S. (2018). Development and characterization studies on magnesium alloy (RZ 5) surface metal matrix composites through friction stir processing. *Journal of Magnesium and Alloys*, *6*(2), 145–163.

Yousefpour, F., Jamaati, R., & Aval, H. J. (2021). Effect of traverse and rotational speeds on microstructure, texture, and mechanical properties of friction stir processed AZ91 alloy. *Materials Characterization*, *178*, 111235.

10 Ultrasonic Welding for Light-Weight Structural Applications
An Industry Perspective

Soumyajit Das and Mantra Prasad Satpathy
KIIT Deemed to be University, Bhubaneswar, India

Ashutosh Pattanaik
JAIN Deemed to be University, Bengaluru, India

Bharat Chandra Routara
KIIT Deemed University, Bhubaneswar, India

Bikash Ranjan Moharana
C. V. Raman Global University, Bhubaneswar, India

CONTENTS

10.1   INTRODUCTION

Aerospace and automotive manufacturers are constantly improving multi-material designs to meet environmental challenges and operational costs (Oliveira, Miranda, and Fernandes, 2017; Li, Zhu, and Guo, 2017; Su et al., 2021; Mohammed et al., 2020). Lightweight structural materials are considered a salient design for improving fuel consumption rate. Due to its satisfactory overall properties and low cost, aluminum alloys have become a leading substitute for a few parts of lightweight applications in terms of economic efficiency (Aonuma and Nakata, 2010; Zhang et al., 2018; de Leon and Shin, 2022). On the other side, Ti alloy has a wide range of structural designs owing to excellent mechanical properties, anti-corrosion and superior formability (Wang et al., 2015; Patel et al., 2019). These lightweight, high performance and low-cost properties are the primary considerations to accept these materials in several industries. However, joining Al-Ti alloy is always a vital aspect for achieving desired quality because of its different physical, mechanical, and chemical properties. The conventional fusion welding process has been typically employed in welding Al and Ti alloy, but it is not an ideal process for joining these two dissimilar metals. For example, Al alloys melt at 660°C while Ti alloys melt at 1650°C (Chen, Quan, and Ke, 2012). This vast difference in the melting points leads to internal stress, massive deformation region and lack of bonding during the welding process (Liu et al., 2014; Zhang et al., 2020; Jasmin et al., 2021). Thus, the mechanical strength is decreased with the increase of intermetallic compounds (IMCs) formation like TiAl, Ti_3Al, $TiAl_3$ and $TiAl_2$ at the weld region as presented in Figure 10.1. To address the

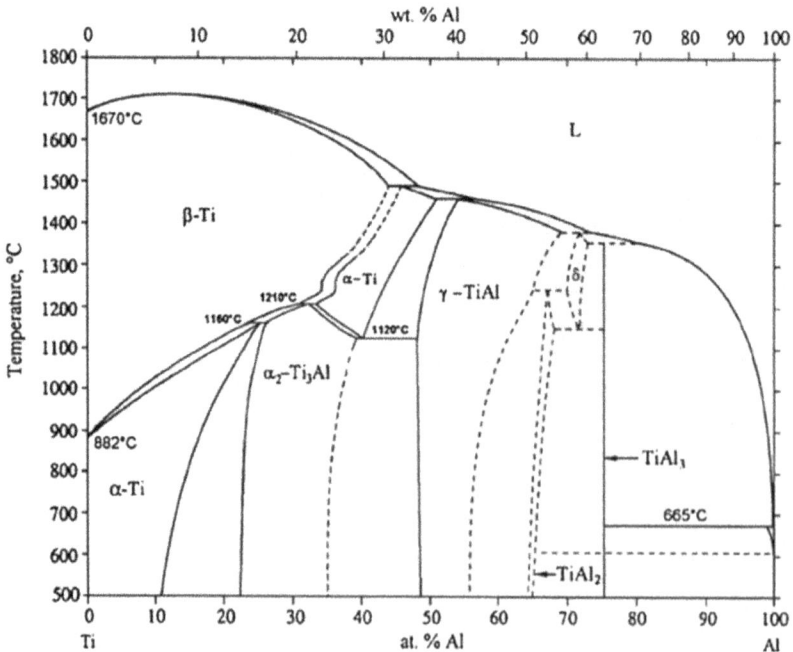

FIGURE 10.1   Al-Ti binary phase diagram (Cui et al., 2012).

aforesaid challenges, ultrasonic welding (USW) is a progressed mechanical methodology to join these Al-Ti alloys under moderate pressure and high-frequency acoustic vibration. As a solid-state joining process, USW prevents melting of the materials and creates a bond formation based on diffusion and adhesion due to interfacial friction. Therefore, it delivers significant weldment strength, and no brittle IMCs formation is noticed around the joint interface (Liu, Cao, and Yang, 2022).

In this comprehensive study, the most frequently used Al-Ti alloys are in-depth investigated on various welding conditions during the USW process. To better understand the effects of weld parameters on the joint characteristics, tensile behavior and fracture mode are explored. In addition, the microstructural characteristics of the weld region between Al-Ti joints are also discussed.

10.2   ULTRASONIC WELDING PROCESS

The USW process is an efficient joining technique that uses high-frequency ultrasonic vibration (20 kHz or higher) to make a bond between the workpiece under the action of clamping pressure. A standard USW machine is based on five fundamental components that consist of an ultrasonic power supply, transducer, booster, sonotrode, and anvil. Electrical energy from the ultrasonic power supply is firstly converted to shear vibration by the transducer. The amplitude of the vibration will then amplify by the booster. The shear vibration is then transmitted to the sonotrode and delivered to the workpiece. The vibrational amplitude is generally represented in the percentage of total ultrasonic power supply and it determines the amount of relative motion and friction at the joint surface. During the USW process, temperature is developed between the workpieces (30–80% of the melting point) and this does not affect the chemical properties of material. It takes only a few seconds to complete this process without the addition of any filler material. Figure 10.2 depicts a schematic illustration of the essential USW equipment.

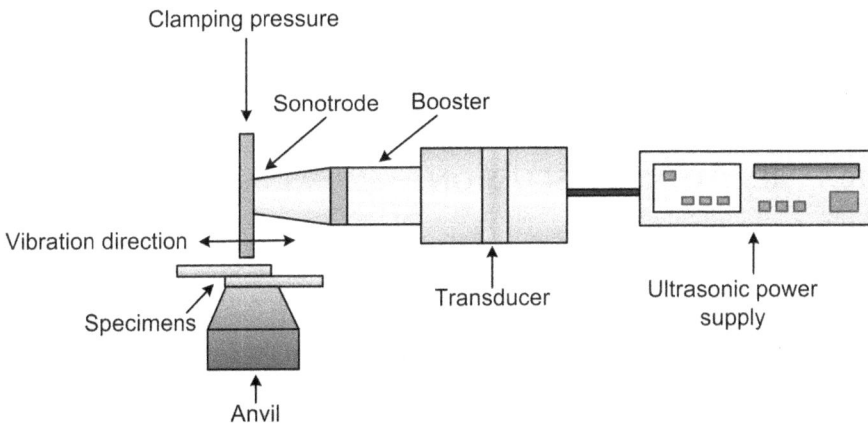

FIGURE 10.2   A schematic illustration of USW equipment.

10.3   DISRUPTIVE TECHNOLOGIES

In recent years, the dynamic market scenario has compelled the aerospace and automotive industries to develop structural integrity of their components. These structural applications involve multi-material design concepts, and a few promising welding processes are evolved to join these dissimilar materials (Su et al., 2021; Liu and Cao, 2021). Moreover, these processes are adopted to increase the bond efficiency and quality of the welded joints. Conventional fusion welding methods for example resistance spot welding (RSW), electromagnetic pulse welding (EPW), explosive welding (EW), various arc welding methods and laser welding have many complications in the dissimilar joining of Al and Ti materials because of their notable changes in thermal and mechanical properties (Jasmin et al., 2021). Solid-state joining methods, such as ultrasonic welding (USW), forge welding (FW), friction stir welding (FSW) and self-piercing rivets, can help to prevent or reduce the degradation of mechanical properties.

10.4   INDUSTRY CHALLENGES

Al and Ti alloys are generally used in automotive parts such as vehicle bodies, steering wheels, stiffeners, brackets, etc. Joining is a crucial phase in these sectors because there are usually 3000–4000 spot welds in a structural body. Therefore, solid-state welding techniques like FSW and USW have gained considerable attention for joining Al-Ti alloys. In order to succeed the overlapping of dissimilar thin sheets, it is difficult to use the FSW process since the stirring pin must rotate and expand into the faying area (Ma et al., 2021). Furthermore, during the FSW process, interface temperature increases beyond a specific range of specimens, and it leads to severe plastic deformation, followed by the crack's formation and distortion in the bond region. Various types of IMCs can be formed during the solidification phase at the joint surface, which usually reduces the bond strength (Zhang et al., 2020; Dhara and Das, 2020). Compared to FSW, the USW technique is usually fast, easy to automate, and can avoid brittle phases IMCs formation at the fusion zone. The heat generation during the USW process is less than the melting point of the material. Therefore, the probability of IMCs formation is radically reduced (Das et al., 2019b). By contrast, USW is more capable of joining thin sheets or multi-layered structural applications in automotive sectors owing to its welding characteristics.

10.5   RESULTS AND DISCUSSION

10.5.1   Mechanical Performance

10.5.1.1   Tensile Shear Strength

Basic validation of the mechanical strength between Al-Ti joints is accomplished by using tensile lap shear test results. Figure 10.3 exhibited the connection of shearing force with different weld times and pressure between Al (AA6061) to Ti (Ti6Al4V) joints. When shearing force is applied to the welded specimens, failure behavior is noticed around the weld interface. This failure mode is indicated as a strong bond formation between Al-Ti joints over the strength of the Al alloy matrix. The shearing

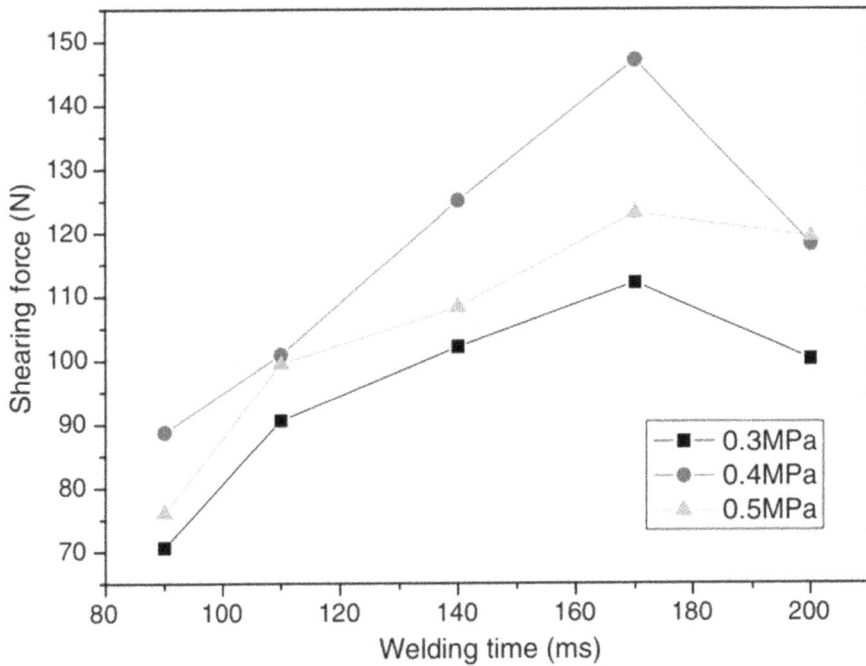

FIGURE 10.3   Weld time vs. shearing force at various weld pressure (Zhu, Lee, and Wang, 2012).

force is initially increased with the respect to weld time under different weld pressure up to a specific limit, and then decreases further due to formation of a residual oxide layer near the weld region. When the weld time is higher, the joint interface temperature is recorded maximum due to the longer vibrational amplitude transmitted for a longer period. It is observed that a higher joint interface temperature can soften the Al surface and deliver a larger amount of plastic deformation at the joint interface, which occurs as a formation of crack in the weld surface (Das, Masters, and Williams, 2019a; Cai et al., 2019). Thus, it is concluded that an ideal weld time always depends on the process parameter, thickness, and properties of the welded material.

10.5.1.2   Micro-Hardness

To measure the hardness value at the joint interface, the Vickers microhardness test is analyzed for further endorsement. Figure 10.4 demonstrated the hardness profile of Al (AA6111) specimen after 30 minutes of welding and four days of natural aging. The variation of hardness values is explained in an almost similar trend. The highest hardness value in the joint surface is noticed after natural aging conditions, which is higher than the value of base metal. Similarly, the lowest hardness value is recorded around the heat-affected zone (HAZ) of the Al surface due to elemental diffusion. After four days of welding, the hardness value is recovered partly owing to the re-precipitation of alloying elements that have been softened during joining (Li and Cao, 2019).

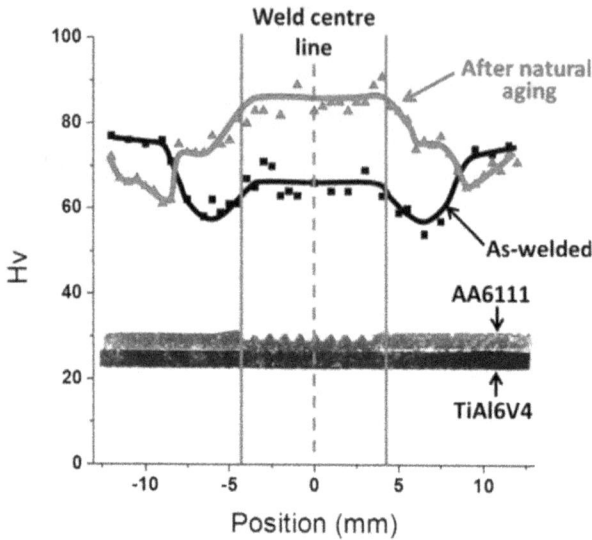

FIGURE 10.4   Hardness values across the Al surface at 30 min (black line) and four days (red line) of Al-Ti welding (Zhang et al., 2014).

10.5.1.3   Fracture Surface Morphology

The effect of sonotrode tip penetration, interface temperature variation, and crack formation around the weld spots are typically observed at various parametric conditions across the welded specimen. The fracture surface of a welded specimen at different weld times is demonstrated in Figure 10.5. When the weld time is minimum, the imprint of the sonotrode tip is significantly less and a slight amount of weld energy is delivered to the joint interface. Due to sliding friction, different sizes of micro-welds are reported at the weld interface of Al-Ti joints. However, the sonotrode tip imprint is deeper with the increases of weld time at the bond surface (Liu, Cao, and Yang, 2022).

10.5.2   Microstructural Analysis

10.5.2.1   SEM and EDS Analysis

To explore the bonded interface more precisely, SEM images are investigated and compared, along with EDS line scan analysis. The SEM images of fracture surface conditions are exhibited in Figure 10.6. Based on the characteristics of the fracture surface on the Al-Ti joint, no obvious difference can be noticed between the mid surface and the end of the nugget region (Figure 10.6a and b). A few small white particles can be seen on the surface of Al (specified by yellow dashed arrows in Figure 10.6c) that are stuffed in Ti and V elements, as confirmed by these results (Figure 10.6e). Small, gray discontinuous 'islands' (shown by red arrows in Figure 10.6d) are detected on the Ti surface, which are mostly enclosed by Al fragments (Figure 10.6f).

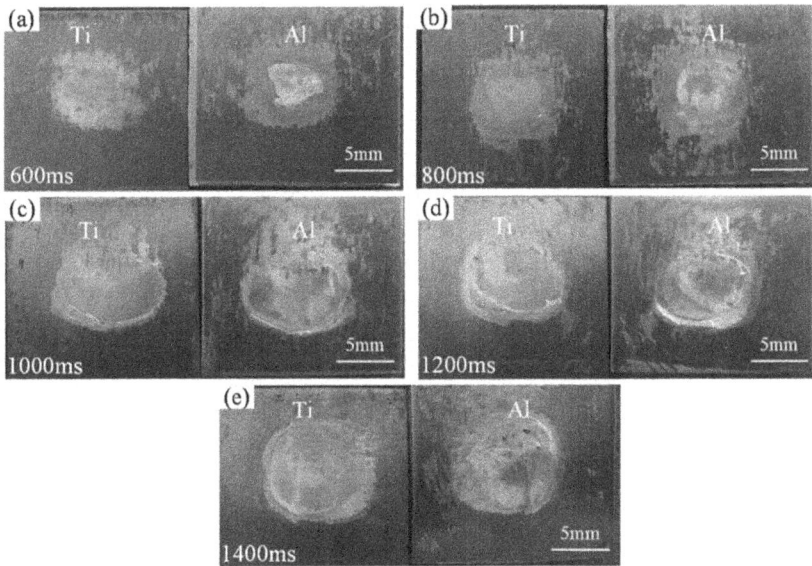

FIGURE 10.5 Fracture surfaces of Al-Ti joints under various weld times (Zhou et al., 2018).

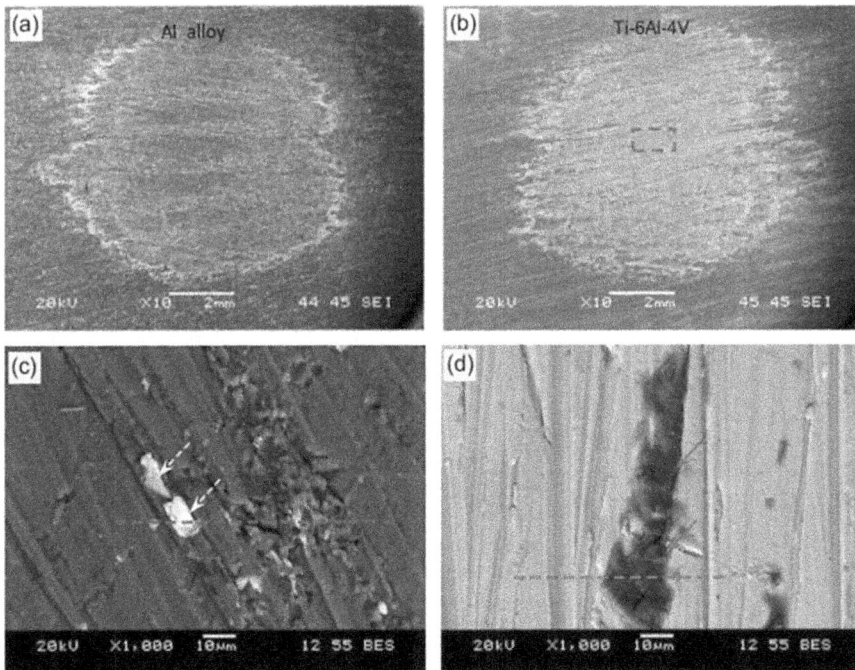

FIGURE 10.6 SEM images of Al-Ti alloy at 1000J of weld energy (a) Al surface (b) Ti surface (c) zoom view (red dash box in Figure 10.6a) of Al surface (d) zoom view (red dash box Figure 10.6b) of Ti surface

(Continued)

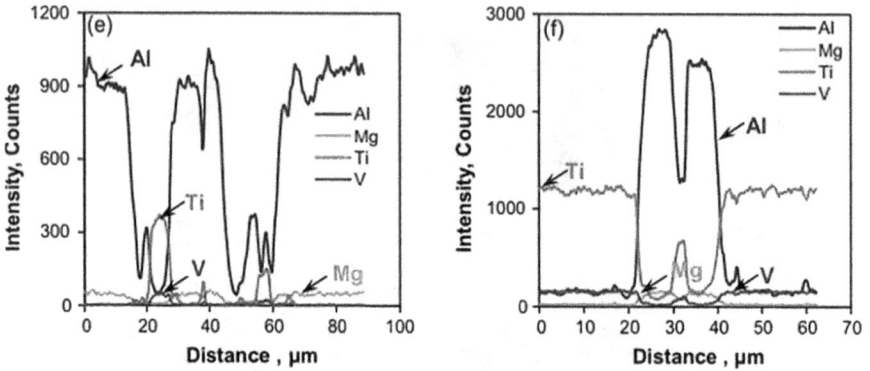

FIGURE 10.6   (Continued)   (e) EDS line scan at the Al interface (indicated by red dash line Figure 10.6c) (f) EDS line scan at the Ti surface (indicated by red dash line Figure 10.6d) (Wang et al., 2015).

The cross-sectional analysis of the bonded specimens at various weld time are disclosed in Figure 10.7. At the weld time of 0.6s, a visible gap is reported in the Al-Ti joints across the bond-line due to an insufficient amount of oxide layer formed between the welded joints (Figure 10.7a and b). When the weld time is 1.4s, the Al-Ti joints are perfectly bonded with each other and make a strong joint formation between these two dissimilar materials (Figure 10.7c). In Figure 10.7d the EDS line scan result disclosed that a small amount of atomic diffusion is noticed with a diffusion distance of around 3 μm at the joint interface These overall results confirmed the absence of IMCs in the weld surfaces, which can influence the final strength of Al-Ti joints (Zhang, Robson, and Prangnell, 2016).

10.5.2.2   XRD and TEM Analysis

The XRD analysis is performed to classify phases of IMCs formation in the joints during the welding process. In these Al-Ti joints, no IMCs formation is spotted on the Al and Ti surface during XRD analysis (Figure 10.8a, b). Even when using the higher resolution TEM images to comprehend the complete microstructures more evidently, no IMCs layer is detected between the Al (AA2139) and Ti (TiAl6V4) specimens. Figure 10.9 shows a bright field and annular dark-field images at the weld interface at various magnifications. Even if there is a layer present, it must be extremely thin (<1 nm), so it is almost negligible. In the comparison of fusion welding and brazing technique, USW is an appropriate process for joining Al-Ti dissimilar alloys to dodge brittle IMCs formation and produce a strong bond.

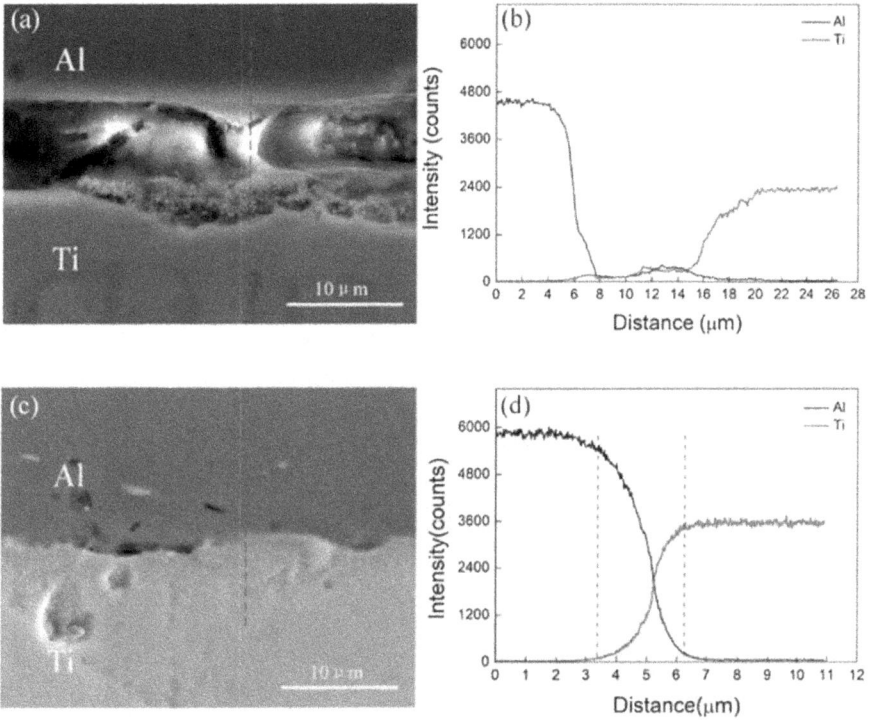

FIGURE 10.7 SEM images of Al-Ti joints at the bond interface (a) weld time of 0.6s (b) EDS line scan at weld time of 0.6s (c) weld time of 1.4s (d) EDS line scan at weld time of 1.4s (Zhou et al., 2018).

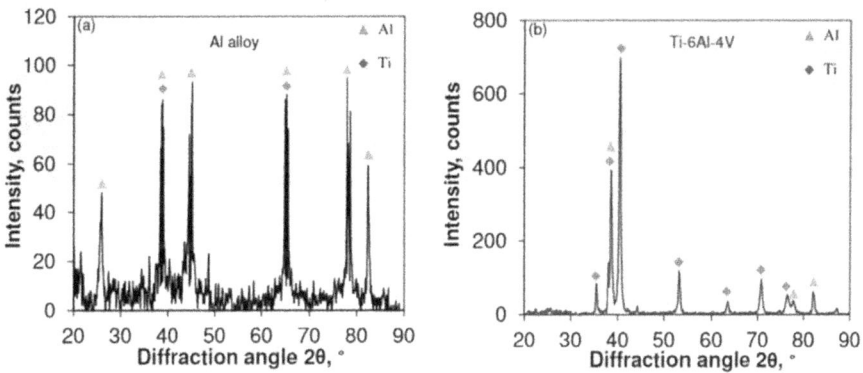

FIGURE 10.8 XRD analysis of Al-Ti dissimilar joints (Wang et al., 2015).

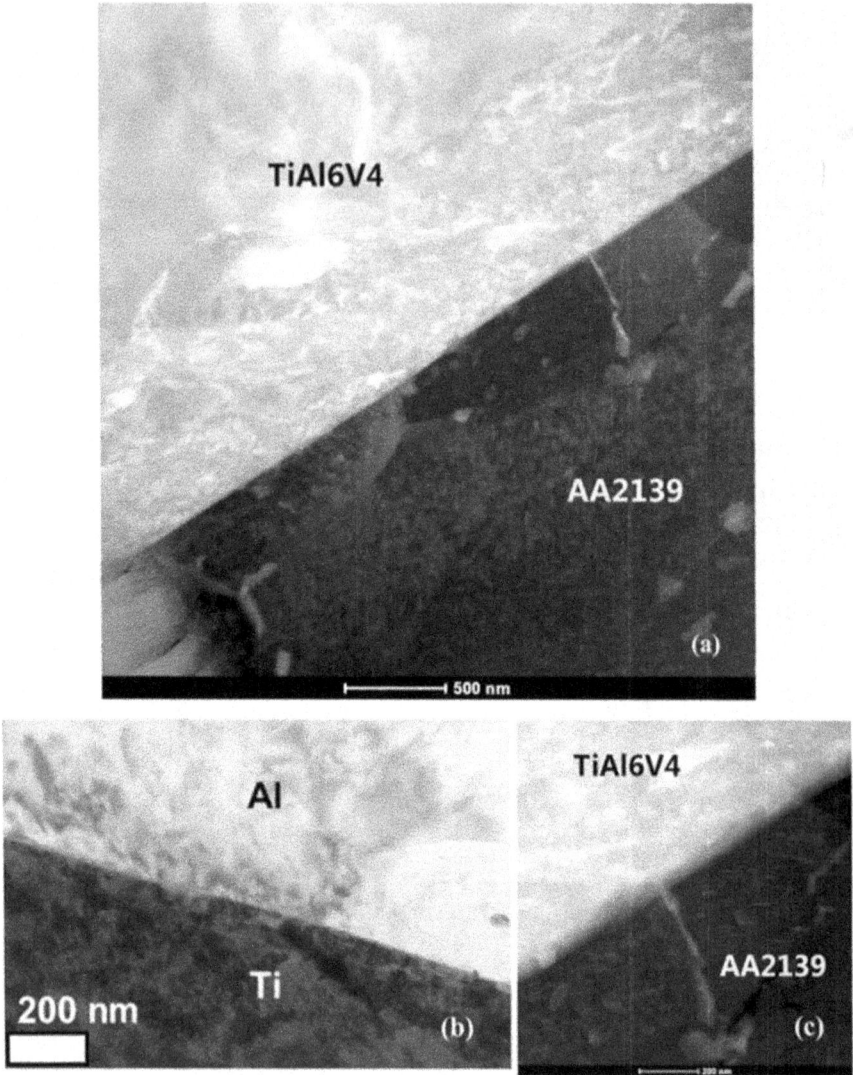

FIGURE 10.9   TEM images (a), weld cross-sectional surface (b), and (c) magnified image of bright field (Zhang, Robson, and Prangnell, 2016).

10.6   CONCLUSIONS

This present study has a comprehensive overview of several bonding phenomena during the USW process. Mechanical performance and microstructural characteristics are investigated to evaluate the weld strength of dissimilar joints. The core conclusions are discussed below:

i. The shearing force increases with the increase of weld time or energy up to a specific limit. After that, this force decreases with increased weld time or energy. Too high or low weld time always reduces the joints' mechanical performance.

ii. The asymmetrical microhardness profile is observed on Al-Ti joints, and no significant changes are noticed on the Ti side compared to the Al surface. The fracture surface is predominantly characterized by different failure modes.

iii. In these Al-Ti joints, it is concluded that variable process parameters and weld conditions can affect the microstructural changes on the bonding surface. From EDS, XRD and TEM analysis, it is confirmed that no IMCs or interfacial layers are generated at the interface of weld spot.

REFERENCES

Aonuma, Masayuki, and Kazuhiro Nakata. 2010. "Effect of Calcium on Intermetallic Compound Layer at Interface of Calcium Added Magnesium–Aluminum Alloy and Titanium Joint by Friction Stir Welding." *Materials Science and Engineering: B* 173 (1–3): 135–38.

Cai, Wayne, Glenn Daehn, Anupam Vivek, Jingjing Li, Haris Khan, Rajiv S Mishra, and Mageshwari Komarasamy. 2019. "A State-of-the-Art Review on Solid-State Metal Joining." *Journal of Manufacturing Science and Engineering* 141 (3): 31012.

Chen, Yu-hua, N I Quan, and Li-ming Ke. 2012. "Interface Characteristic of Friction Stir Welding Lap Joints of Ti/Al Dissimilar Alloys." *Transactions of Nonferrous Metals Society of China* 22 (2): 299–304.

Cui, Xiping, Guohua Fan, Lin Geng, Yin Wang, Lujun Huang, and Hua-Xin Peng. 2012. "Growth Kinetics of TiAl3 Layer in Multi-Laminated Ti--(TiB2/Al) Composite Sheets during Annealing Treatment." *Materials Science and Engineering: A* 539: 337–43.

Das, Abhishek, Iain Masters, and David Williams. 2019a. "Process Robustness and Strength Analysis of Multi-Layered Dissimilar Joints Using Ultrasonic Metal Welding." *The International Journal of Advanced Manufacturing Technology* 101 (1–4): 881–900.

Das, Soumyajit, Mantra Prasad Satpathy, Ashutosh Pattanaik, and Bharat Chandra Routara. 2019b. "Experimental Investigation on Ultrasonic Spot Welding of Aluminum-Cupronickel Sheets under Different Parametric Conditions." *Materials and Manufacturing Processes*, 34 (15): 1689–170.

Dhara, Sisir, and Abhishek Das. 2020. "Impact of Ultrasonic Welding on Multi-Layered Al--Cu Joint for Electric Vehicle Battery Applications: A Layer-Wise Microstructural Analysis." *Materials Science and Engineering: A* 791: 139795.

Jasmin, N. Mary, K. Anton Savio Lewise, D. Raguraman, M. Murugan, S. S. V. S. Aditya, T. S. Abhishek Reddy, S. Hari Krishna, and Ram Subbiah. 2021. "An Overview on Characteristics and Performance of Ultrasonic Welding Process on Different Materials." *Materials Today: Proceedings*, 50 (5): 1508–1510.

de Leon, Michael, and Hyung-Seop Shin. 2022. "Review of the Advancements in Aluminum and Copper Ultrasonic Welding in Electric Vehicles and Superconductor Applications." *Journal of Materials Processing Technology*, 307: 117691.

Li, H., and B. Cao. 2019. "Effects of Welding Pressure on High-Power Ultrasonic Spot Welding of Cu/Al Dissimilar Metals." *Journal of Manufacturing Processes* 46: 194–203.

Li, Qiao, Yuanxiang Zhu, and Jialin Guo. 2017. "Microstructure and Mechanical Properties of Resistance-Welded NiTi/Stainless Steel Joints." *Journal of Materials Processing Technology* 249: 538–48.

Liu, Jian, and Biao Cao. 2021. "Microstructure Characteristics and Mechanical Properties of the Cu/Al Dissimilar Joints by Electric Current Assisted Ultrasonic Welding." *Journal of Materials Processing Technology* 297: 117239.

Liu, Jian, Biao Cao, and Jingwei Yang. 2022. "Texture and Intermetallic Compounds of the Cu/Al Dissimilar Joints by High Power Ultrasonic Welding." *Journal of Manufacturing Processes* 76: 34–45.

Liu, Kun, Yajiang Li, Shouzheng Wei, and Juan Wang. 2014. "Interfacial Microstructural Characterization of Ti/Al Joints by Gas Tungsten Arc Welding." *Materials and Manufacturing Processes* 29 (8): 969–74.

Ma, Qiuchen, Yichen Cao, Wenwu Zhang, Weiwei Zhao, Hongtao Chen, Mingyu Li, Zhimin Liang, Yong Xiao, and Hongjun Ji. 2021. "Low Energy Ultrasonic Welding for Cu-Cu Joining Accelerated via Cu Nanoparticles." *Journal of Materials Processing Technology* 296: 117210.

Mohammed, S. M. A. K., Y. D. Jaya, A. Albedah, X. Q. Jiang, D. Y. Li, and D. L. Chen. 2020. "Ultrasonic Spot Welding of a Clad 7075 Aluminum Alloy: Strength and Fatigue Life." *International Journal of Fatigue* 141: 105869.

Oliveira, J. P., R. M. Miranda, and F. M. Braz Fernandes. 2017. "Welding and Joining of NiTi Shape Memory Alloys: A Review." *Progress in Materials Science* 88: 412–66.

Patel, V. K., S. D. Bhole, D. L. Chen, D. R. Ni, B. L. Xiao, and Z. Y. Ma. 2019. "Ultrasonic Spot Welding of Dissimilar 2024Al Alloy and SiCp/2009Al Composite." *Proceedings of the Institution of Mechanical Engineers, Part L: Journal of Materials: Design and Applications* 233 (3): 531–38.

Su, Zhanzhan, Zhengqiang Zhu, Yifu Zhang, and Hua Zhang. 2021. "Analysis of Microstructure and Mechanical Properties of AZ31B Magnesium Alloy/AA6061 Aluminum Alloy Welded Joint by Ultrasonic Welding." *Materials Research* 24 (2): 1–9.

Wang, S. Q., V. K. Patel, S. D. Bhole, G. D. Wen, and D. L. Chen. 2015. "Microstructure and Mechanical Properties of Ultrasonic Spot Welded Al/Ti Alloy Joints." *Materials & Design* 78: 33–41.

Zhang, C. Q., J. D. Robson, Octav Ciuca, and Philip B. Prangnell. 2014. "Microstructural Characterization and Mechanical Properties of High Power Ultrasonic Spot Welded Aluminum Alloy AA6111--TiAl6V4 Dissimilar Joints." *Materials Characterization* 97: 83–91.

Zhang, C. Q., J. D. Robson, and P. B. Prangnell. 2016. "Dissimilar Ultrasonic Spot Welding of Aerospace Aluminum Alloy AA2139 to Titanium Alloy TiAl6V4." *Journal of Materials Processing Technology* 231: 382–88.

Zhang, Wei, S. S. Ao, Joao Pedro Oliveira, Zhi Zeng, Zhen Luo, and Z. Z. Hao. 2018. "Effect of Ultrasonic Spot Welding on the Mechanical Behaviour of NiTi Shape Memory Alloys." *Smart Materials and Structures* 27 (8): 85020.

Zhang, Wei, Sansan Ao, J. P. Oliveira, Chunjie Li, Zhi Zeng, Anqi Wang, and Zhen Luo. 2020. "On the Metallurgical Joining Mechanism during Ultrasonic Spot Welding of NiTi Using a Cu Interlayer." *Scripta Materialia* 178: 414–17.

Zhou, L., J. Min, W. X. He, Y. X. Huang, and X. G. Song. 2018. "Effect of Welding Time on Microstructure and Mechanical Properties of Al-Ti Ultrasonic Spot Welds." *Journal of Manufacturing Processes* 33: 64–73.

Zhu, Zhengqiang, Kang Yong Lee, and Xiaolong Wang. 2012. "Ultrasonic Welding of Dissimilar Metals, AA6061 and Ti6Al4V." *The International Journal of Advanced Manufacturing Technology* 59 (5–8): 569–74.

11 Supervised Machine Learning Algorithms for Machinability Assessment of Graphene Reinforced Aluminium Metal Matrix Composites

Suryakant Muduli and Trupti R. Mahapatra
Veer Surendra University of Technology, Odisha, India

Allu V. K. Murty
AIML Architect, Bengaluru, India

Soumya R. Parimanik and Debadutta Mishra
Veer Surendra University of Technology, Odisha, India

Pratap Chandra Padhi
CIPET: Institute of Petrochemicals Technology (IPT),
Bhubaneswar, Odisha, India

CONTENTS

DOI: 10.1201/9781003346623-11

11.1 INTRODUCTION

Aluminum has been popularly used in automobile and aircraft engine components and has replaced the use of heavy metal iron owing to the reduction in weight of the structure/structural component. However, pure aluminum alone does not own all the essential properties, such as high specific strength, good wear resistance, well-regulated thermal expansion coefficient, and low density with high conductivity (Behera et al., 2020). Normally, micro-sized or nano-level particles are used in the preparation of Aluminum Metal Matrix Composite (AMMC) with an intent to improve the tensile strength and the yield strength of the composites, but at the cost of ductility. A variety of combinations of matrixes and reinforcement materials have been tested and verified. Graphene (first patented by Nobel Laureates Konstantin Novoselov and Andre Geim in the year 2002), one of the carbon allotropes having strong covalent bonds between carbon atoms, has the highest tensile strength of 130.5 GPa and Young's modulus of 1TPa (40 times stronger than diamond and 300 times stronger than structural steel). It has a high melting point (appx. $4125°K$), and excellent thermal and electrical conductivity at room temperature. The biggest advantage of graphene to be propitiously used as the filler material is the high strength to weight ratio. The mass of one layer of 1 m^2 area is just 0.77 mg (The Royal's Academy of Sciences [The Nobel Prize in Physics], 2010). Thus, promising graphene has been used as a filler in AMMCs to enhance the desired properties for targeted applications (Khanna et al., 2021). Different production methods are implemented to synthesize the MMCs based on the matrix material. For hybrid AMMCs, liquid route (reactive processing, spray deposition, infiltration casting, stir casting, reactive processing) techniques and solid route (vapor deposition techniques, diffusion bonding, and powder metallurgy) techniques are adopted (Bodunrin et al., 2015). In the former methods, like the stir casting/compo casting, a comparatively low-cost process, the particulate fibers are added to the molten metal, and the molten metal is continuously stirred by mechanical means during solidification. One of the major demerits of this process is that homogeneity is not proper because of the accumulation of suspended particles. However, this inhomogeneity can be minimized by performing proper stirring (Suthar and Patel, 2018). The stir casting has been recommended to be the most adaptable and economical process for the synthesis of the AMMCs at the industrial level (Venkatesan and Xavior, 2018). However, inadequate particle distribution, porosity, and wettability are the key worries associated with it, which can be minimized by suitably performing the stirring operation (Suthar and Patel, 2018).

Industry 4.0 uses cyber-physical systems (for process monitoring), the Internet of things (IoT) (real-time communication of different systems with each other alongside collaboration with other human systems and operators), cloud computing (central storage of data logged by cyber-physical systems), and machine learning (ML) (extraction of the existing patterns in the data). Due to the advancements in ML, researchers attempt to utilize the same in machining processes to improve productivity alongside product quality, monitor the health of the system with adequate equipment availability, and to optimize the process as well as the design parameters leading to a new machining paradigm referred as smart manufacturing (Kim et al., 2018). Thus, we note the implementation of ML in several frameworks of machining processes, from tool condition monitoring (Caggiano, 2018), energy consumptions (Brillinger et al., 2021), and lowering product costs to surface quality predictions (Aggogeri et al., 2021), in comparison to the existing and modified soft-computing techniques (He et al., 2019). Radhika et al., 2013) built a classifier model based on the random forest (RF) algorithm for the prediction of surface roughness (R_a) in the turning of hybrid MMCs using the vibration signal of the machine tool capable of detecting the condition automatically. A comparative analysis for the prediction ability of cutting parameters (cutting force and R_a) in high-speed turning (Jurkovic et al., 2018) revealed the superiority of polynomial regression over the support vector regression (SVR) and the artificial neural network (ANN). A coupled adaptive neuro-fuzzy inference system (ANFIS) and quantum particle swarm optimization (QPSO) machine learning approach has been implemented (Alajmi and Almeshal, 2020) for the prediction of R_a of AISI304 stainless steel during the dry and cryogenic turning process. The approach not only depicted faster convergence and robustness, but also resulted in greater accuracy in comparison to ANFIS, ANFIS-GA, ANFIS-PSO when the experimentally acquired dataset is used for training and testing of the model. A Bayesian theory based Gaussian process regression (GPR) ML method (Alajmi and Almeshal, 2021) also portrayed greater accuracy over SVM and ANN in predicting the cutting force and optimizing the process parameters in the turning of AISI4340 alloy steel. The performance of the GPR is more robust and accurate in predicting the Ra as compared to the ANN in turnings of brass metals (Zhang and Xu, 2022). Moreover, it has been revealed that, though the polynomial regression (PR) is faster in terms of performance, the accuracy of SVM and GPR is greater than the PR in predicting and optimizing the cutting forces as well as R_a (Singh et al., 2022). In other research (Dubey et al., 2022a), the linear regression (LR) outperformed the random forest regression (RFR) and the SVM in predicting the cutting force while machining AISI304 steel in a mist lubrication environment. However, the AdaBoost regression (ABR) and gradient boost regression (GBR) are reported to be poorer than the PR and the RFR for predicting the output responses during hard machining of AISI D6 steel (Das et al., 2022). The performance of the XGBoost regression (XGBR) model is informed to be better than the RFR and the decision tree regression (DTR) model in predicting the machining error of the turning process (Wang, Kuo, and Chen, 2022). The ML predictive model developed using the feed forward neural network (FFNN) is found to be more accurate in comparison to the traditional regression model for predicting the Ra in drilling of Nimonic C263 (Lakshmana Kumar et al., 2022). Further, the ARX441 and ARMAX3331 have been proposed to be the best suited predictive models for the micro and nano MMCs, respectively (Sekhar et al., 2022). When the LR, RFR, and SVM have been utilized for estimating the Ra values in minimum quantity lubrication (MQL) assisted turning of AISI304 steel (Dubey et al., 2022b), the RFR outperformed the other two with fewer errors and a higher R-score. The case studies (Shanmugasundar et al., 2021) on prediction ability of the LR, RFR, and ABR in electro discharge machining (EDM) suggested the RFR and ABR to be more suitable. Moreover, diverse simple and hybrid ML techniques have been proposed and implemented in the recent past for monitoring the tool wear (de Farias et al., 2020; Tabaszewski et al., 2022) and predicting the residual stresses (Elsheikh et al., 2021) during the turning operation.

It is imperative from the review of literature that though graphene has been hailed as a superior material in the manufacturing domain in particular, inadequate studies were reported and extensive information is essential regarding the machining and machinability of graphene reinforced MMCs. Moreover, the ML is a versatile technique that can be implemented towards quality monitoring and prediction for assuring high-quality production (Misaka et al., 2020; Cheng et al., 2020). However, the model performance in terms of robustness, stability, and accuracy are the key measures to choose among diverse methods (Du et al., 2021; Ulas et al., 2020). In the present work, three different supervised ML algorithms are implemented for the prediction of the rate of material removal (RMR), and R_a values during the turning graphene reinforced AMMCs and their performances are compared with the results achieved via the back propagation ANN model. Cylindrical shaped samples of graphene (0, 0.5, 1, 2, and 3 wt.%) reinforced MMCs with pure aluminum as matrices are prepared by the stir casting method. Subsequently, a dry turning operation is carried out on a CNC lathe using a PCD carbide insert tool, according to full factorial design. The machining performance indicators (RMR and R_a) are acquired by considering the wt.% of graphene alongside the speed, feed, and depth of cut as the controllable factors. These data are utilized as input to train, validate, and subsequently test several well-established ML regression models to bring out insights regarding the performance of each model in predicting the desired responses and validated via confirmation runs.

11.2   METHODOLOGY AND RESULTS

In the present study, Al is used as the metal matrix material because it is abundant in mines, cheaper, and owing to its corrosion resistance, lightweight. It is widely used in MMC along with silicon and boron oxides. Moreover, owing to their enhanced mechanical properties as well as oxidation resistance, electrical/thermal conductivity, and biocompatibility, graphene reinforced MMCs have received a lot of attention (Xiang et al., 2016). So, as far as the MMC manufacturing methods are concerned, the stir casting method (that involves producing the melt of the matrix material and then reinforcing the selected material through appropriate dispersion by stirring) is the most economical. In order to study the machinability of Al/graphene MMC, cylindrical specimens having a diameter 25 mm and length of 120 mm with different weight percentages of graphene reinforcement (0, 0.5, 1, 2 and 3 wt.%) are prepared. In the present stir casting process, the graphene wt.% is mixed with a specified amount of aluminum powder with the help of a stirrer, put in a crucible and heated to the melting point of the aluminum (700°C) inside the furnace in a nitrogen gas environment. Near net shape is achieved by preparing mold with over tolerance to accommodate the shrinkage effect. Finally, the molten metal with graphene reinforcement is continuously stirred by mechanical means (600 rpm), put inside the mold, allowed to solidify, and broke open to get the required specimen.

The graphene particulates are abrasive and act like a cutting edge, resulting in fast tool wear and vibration in the machine tool, and subsequently, affect the machinability of the AMMCs. Therefore, the impact of basic machining parameters on the machinability of MMCs with and without graphene reinforcement must be critically assessed. A CNC lathe machine of MTAB make is selected to perform the turning operation on the currently prepared Al/graphene MMCs. The insert DCGT11T304 NS LT05 that can be used in the above CNC lathe is a carbide insert. For initial dry cut, it has been planned to cut with carbide then substitute it with either PCD or CBN inserts as the CBN/PCD perform better than carbide insert and can be operated at high cutting speed. The PCD inserts used in the present analysis are ALUMINA make with tool geometry DCMW 11T 302 PD 103 with nose radius 0.2 mm. The tool holder specification is SDSCL-121211-F3, and accordingly the insert has been selected. The brazed part of the PCD tip with the carbide base pictorial view and SEM image of the tip is shown in Figure 11.1.

PCD insert and SEM image of PCD tip. Here the brazed part of the PCD tip with the carbide base pictorial view and SEM image of the tip is shown.

Generally, for turning operations depth of cut, feed, and speed are the normal cutting parameters. In this present research, a facing operation is performed with a different feed, width of cut (WOC), and RPM. The WOC ranges from 0.1 mm to 0.2 mm. As the nose radius of the insert is 0.2 mm, the optimized DOC is 1/3rd of the nose radius. However, from the research point of view, 0.2 mm maximum WOC is taken. Instead of speed as one of the controlled parameters, RPM is chosen, as the speed would vary with the change in diameter during facing operation. Moreover, in the control panel and programming, RPM is controlled but not the speed. As the maximum spindle RPM can be achieved by the existing CNC lathe is 3000, the range of RPM for experimentation is selected as 1000–3000 RPM. Finally, the feed range is chosen as 100–200 mm/min as per the recommendation of the tool supplier and the literature review. In addition, the wt.% of graphene is taken as a controllable input parameter that varies as 1–3 wt.%. The details of the cutting parameters and their levels have been summarized in Table 11.1.

In the present investigation, MRR in gm/sec and roughness average (R_a) in μm have been considered as the output response characteristics. For better machinability and productivity, MRR should be high. For a quality surface finish, the R_a values should be low. A digital weighing machine with least count (LC) 0.0001 g and a Mitutoyo SJ210 Surface Roughness tester have been used for acquiring the desired output responses. For a systematic approach to conduct the experiment, a

FIGURE 11.1 PCD insert and SEM image of PCD tip.

TABLE 11.1
Cutting Parameters and Their Levels

Level	Cutting Speed (S) (rpm)	Feed Rate (F) (mm/min)	WOC (D) (mm)	Wt.% Graphene (G) (wt.%)
1	1000	100	0.1	0
2	2000	150	0.15	0.5
3	3000	200	0.2	1.0
4	–	–	–	2.0
5	–	–	–	3.0

full factorial design of experiment (DOE) is followed. Thus, 135 dry turning experiments with different combinations of controllable parameters are conducted to acquire desired responses (RMR and Ra values) and listed in Table 11.2. Now, this dataset has been subsequently used as input for ANN and various supervised ML regression models for assessing their performance and efficacy to predict of the RMR and R_a values and subsequently perceive the best performing metamodel. In order to do that, the sequence of steps as depicted in Figure 11.2 has been followed.

Flow chart showing sequence of steps in the present ML regression models. 135 dry turning experiments with different combinations of controllable parameters are conducted to acquire desired responses (RMR and R_a values). This dataset has been subsequently used as input for ANN and various supervised ML regression models. The sequence of steps followed to obtain the best performing metamodel is depicted in the figure.

Step 1: (Import Libraries): In this step, all the required libraries that would be needed at various steps of the program for various models are imported.

Step 2: (Data Collection): The data frames are created using the numerically prepared dataset to hold the input data for various models.

Step 3: (Transform): Post data collection, various transformation or data enrichment is performed viz. label encoding (for discreet values), train/test split, scaling, and normalizing the data as required. It is extremely needed as per industrial best practices for highly efficient model outcomes.

Step 4: Design, build, and tune model: This step is an iterative step, along with the tuning of the model for various hyperparameters so that the learning rate of the respective model is high, which reflects the best "R^2" value obtained.

First, necessary libraries are imported to run the main program. Pandas is used for data frame creation and handling. Numpy is used for numerical analysis multidimensional array handling. The sci-kit learn module provides various submodules to pre-process the data, create and train the machine learning model, and to test its performance. Matplotlib is used for the plotting of graphs.

TABLE 11.2
Experimentally Acquired Dataset [RMR and R$_a$]

Feed (m/min)	RPM (rev/min)	DOC (mm)	0 wt.% Graphene		0.5 wt.% Graphene		1 wt.% Graphene		2 wt.% Graphene		3 wt.% Graphene	
			RMR(gm/s)	R$_a$(μm)	RMR(gm/s)	R$_a$(μm)	RMR(gm/s)	R$_a$(μm)	RMR(gm/s)	R$_a$(μm)	RMR(gm/s)	R$_a$(μm)
100	1000	0.10	0.01270	0.418	0.00270	1.761	0.01425	1.070	0.01258	0.686	0.01283	0.499
100	1000	0.15	0.02033	0.959	0.01803	0.390	0.02003	0.745	0.01975	0.708	0.01947	0.414
100	1000	0.20	0.03100	0.331	0.02770	0.780	0.02386	0.769	0.02682	0.925	0.02599	0.629
100	2000	0.10	0.01272	0.378	0.00913	0.989	0.01675	0.713	0.01329	0.88	0.01178	0.674
100	2000	0.15	0.01991	0.441	0.02501	0.571	0.01924	0.757	0.02144	1.743	0.01617	0.572
100	2000	0.20	0.02593	0.811	0.02775	0.644	0.02527	1.250	0.02250	2.925	0.02711	0.511
100	3000	0.10	0.01204	1.001	0.01215	0.442	0.01460	0.997	0.01274	0.979	0.01171	1.406
100	3000	0.15	0.01964	0.415	0.02275	0.584	0.01918	0.936	0.01884	0.455	0.02007	0.718
100	3000	0.20	0.02658	0.429	0.02778	0.729	0.02468	0.331	0.02762	0.746	0.02346	0.622
150	1000	0.10	0.02086	0.316	0.01185	1.826	0.02150	2.200	0.01580	0.972	0.02198	1.001
150	1000	0.15	0.03148	0.517	0.03061	1.500	0.02700	1.181	0.02754	1.243	0.02776	0.704
150	1000	0.20	0.04184	0.355	0.03950	2.103	0.03809	1.202	0.04125	2.609	0.03694	0.810
150	2000	0.10	0.01919	0.713	0.01781	1.456	0.02113	1.050	0.01653	1.005	0.01961	0.593
150	2000	0.15	0.02950	0.674	0.03206	0.980	0.02384	0.632	0.03044	0.775	0.02808	0.698
150	2000	0.20	0.04077	0.428	0.03491	0.763	0.04031	0.708	0.03600	0.665	0.03694	0.611
150	3000	0.10	0.01890	0.342	0.02155	1.849	0.02052	1.134	0.02222	1.057	0.01934	0.779
150	3000	0.15	0.03073	0.290	0.03325	2.264	0.03015	0.926	0.02976	0.792	0.02823	0.910
150	3000	0.20	0.04081	0.327	0.03982	0.394	0.04132	0.931	0.03705	0.537	0.04166	0.772
200	1000	0.10	0.02145	0.635	0.02782	2.815	0.01559	1.694	0.02387	2.594	0.02007	1.376
200	1000	0.15	0.04074	0.492	0.04697	3.236	0.04228	2.154	0.03838	1.989	0.03370	0.996
200	1000	0.20	0.05181	0.407	0.05002	2.764	0.05235	1.387	0.05421	3.853	0.04998	1.236
200	2000	0.10	0.02293	0.390	0.02286	0.692	0.01810	0.410	0.01615	0.468	0.02314	0.647
200	2000	0.15	0.02500	0.446	0.03403	0.881	0.03861	0.835	0.03750	1.504	0.04111	0.708
200	2000	0.20	0.04958	0.381	0.05072	0.617	0.04667	0.729	0.04441	0.695	0.05060	0.586
200	3000	0.10	0.02299	0.353	0.02869	0.782	0.02249	1.024	0.02522	0.629	0.02191	0.906
200	3000	0.15	0.03739	0.356	0.02957	0.945	0.03383	1.025	0.03314	0.783	0.03893	0.927
200	3000	0.20	0.05376	0.452	0.07428	0.528	0.05429	1.045	0.04859	0.593	0.04725	1.284

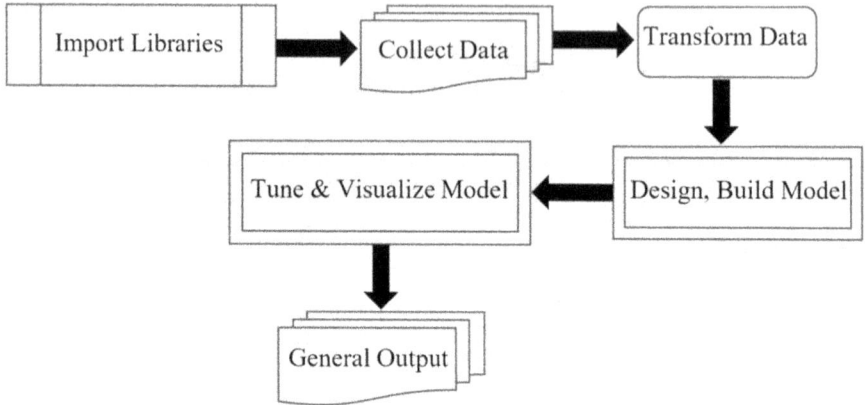

FIGURE 11.2 Flow chart showing sequence of steps in the present ML regression models.

11.3 IMPLEMENTATION OF ML

Following are the four types of ML regression models that are implemented in the present study and compared to the respective performance metric. For the data pre-processing, analysis, and subsequent prediction, the NumPy, scikit-learn library, and pandas libraries of the python language have been utilized. In each case, the data is split into two parts, i.e., 80% data for the training purpose and 20% data for the testing purpose.

Artificial Neural networks (ANN) model: These computational algorithms are envisioned to simulate the behaviour of biological systems encompassing "neurons" connected by arcs. It corresponds to dendrites and synapses. These systems of unified "neurons" compute values from inputs. Each arc is associated with a weight at each node. The input values for the nodes and activation function along the incoming arcs are defined alongside the adjustment by the weights of the arcs. A presently implemented ANN model contains the following 3 layers as depicted in Figure 11.3.

Input layer – The raw information fed into the network are represented by the activity of the input layer.

Hidden layer – The activity of each hidden unit represents the weights applied to the inputs and performs nonlinear transformation by leading them as the output by means of an activation function.

Output layer – The activity of the hidden units and the weights assigned governs the behaviour of the output units.

Schematic representation of ANN model. The 3 layers of presently implemented ANN is shown in the figure. The performance accuracy of the model in terms of the Ra value prediction is elaborated.

In the present analysis, the MLPRegressor() function is used to initiate a neural network model object. The *fit* function is used to train the model using the train datasets given within them, and score function is used to test the performance of the model. Various permutations, and combinations of input, and/or hidden layers with several hyperparameters have been accomplished and lastly, the following combinations as provided in Table 11.3 are attained that resulted R^2_score (R^2 value) of 0.833.

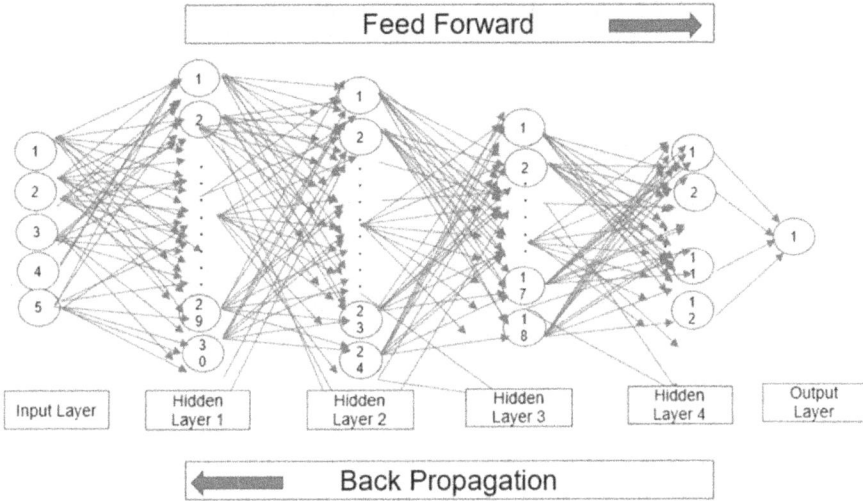

FIGURE 11.3   Schematic representation of ANN model.

TABLE 11.3
Ultimate Input, Hidden Layers and Hyperparameters of the ANN Model

Sl. No.	Parameters	ANN Values
1	No of Input Layer Units	5
2	No of Hidden Layer	4
3	No of 1st Hidden Layer Units	30
4	No of 2nd Hidden Layer Units	24
5	No of 3rd Hidden Layer Units	18
6	No of 4th Hidden Layer Units	12
7	No of Output Layer Units	1
8	No of Epochs	600
9	Batch Size	10

Moreover, alternative supervised ML (SML) methodologies, such as the Decision Tree Regression (DTR) Model, Random Forest Regression (RFR) Model, and Light Gradient Boost Regression (LGBR) Model have been implemented to create computational systems that learn from data for predictions and inferences. As the data is labeled well, we can apply the SML to teach or train the machine. Choosing the best algorithm for the given dataset and problem out of various existing algorithms in ML is vital while creating a ML model.

a. Decision Tree Regression (DTR) Model: DTR Model is a supervised regression technique used for both classification and regression problems. It correlates the target column and the independent variables and express them as a tree structure, where internal nodes, branches, and each leaf node represents the features of a dataset, the decision rules, and the outcome, respectively. It is accomplished by binary splitting data, using functions based on comparison operators on the independent columns. Algorithms for constructing decision trees usually work top down, by choosing a variable at each step that best splits the set of items, that equally works well with both categorical and quantitative variables. The *tree. DecisionTreeRegressor()* function is used to initiate a decision tree model object. The *fit* function is used to train the model by using the train datasets given within them, and the *score* function is used to test the performance of the model. With multiple combinations of hyperparameters alongside pruning, a suitable train-test split is chosen. The *GridsearchCV* function creates a *gridsearch* object that takes another model object, parameter grid, and a number of cross validations to be done as input. Upon fitting the *gridsearch* model, an object with train data, it trains and validates itself to find the best set of parameters from the given parameter grid. It is worth mentioning that the *GridsearchCV* uses a cross validation algorithm to split the train data into a particular number of parts and uses one part to validate the model and other parts to train the model. At last, it can find the highest scoring model and its parameters. Out of various train_test splits, we considered the best fit model for the combination at which the R^2_Score_Train=1.0 and R^2_Score_Test=0.813. It is clearly observed that there is considerable difference in R^2_Score from train and test datasets, indicating the sign of overfitting in the resultant DTR models.

b. Random Forest Regression (RFR) Model: It is an SML algorithm used for Classification as well as Regression problems in ML and is used on a labeled dataset in regression problems. It generates decision trees on a randomly selected subsample of the complete dataset, calculates prediction on each decision tree, and selects the best decision tree by the means of voting. RFR is used in various domains, such as feature selection and recommendation engines. It lays the foundation of the Boruta algorithm, which is used in the selection of important features in the dataset. It has the advantage of handling large datasets with high dimensionality and is known for preventing the overfitting issue, resulting in greater accuracy. In this analysis, the *RandomForestRegresson()* function is used to initiate the model object, the *fit* function is used to train the model using the train datasets given within them, and then the *score* function is used to test the performance of the model. The *GridsearchCV* streamlines the process of finding the best sets of parameters for a given model and is provided when a parameter grid and a basic model is given to it. With multiple combinations of hyperparameters and a train-test split, the best R^2_Score_Train and R^2_Score_Test values were observed as 0.979 and 0.854, respectively and chosen as the best fit model.

c. Light Gradient Boost Regression (LGBR) Model: LightGBMR splits the tree leaf-wise contrary to other boosting algorithms like those grown tree level-wise. The prediction model is created in the form of an ensemble of weak prediction models, typically decision trees. The leaf with maximum delta loss is chosen to grow. The leaf-wise algorithm has lower loss than the level-wise algorithm as the leaf is fixed. However, the complexity of the model may be increased and lead to overfitting in small datasets. An implicit feature selection method and irrepressibility to overfitting makes the model popular for regression problems. However, it is associated with difficulties in scaling the algorithm, and is slower to train. Here, a basic gradient boosting model is created, trained, and tested. *GradientBoostingRegressor()* function is used to initiate a gradient boosting model object, *fit* function is used to train the model using the train datasets given within them and the *score* function is used to test the performance of the model. With multiple combinations of hyperparameters and train-test splits for the present dataset, the combination with R^2_Score_Train value 0.974 and R^2_Score_Test value 0.892 is considered as the best fit model and utilized for the prediction purpose.

11.4 DISCUSSION OF THE RESULTS

As mentioned previously, for evaluating the performance of the different supervised ML techniques, the dataset is prepared by performing a dry turning operation according to the full factorial DOE with four parameters (cutting speed, feed, WOC, and wt.% of graphene) as the input and two target columns, i.e., the RMR and Ra values. The total 135 observations in dataset are divided into train and test dataset categories. The training data is used to train the models, and the testing data is used to validate the deducement of the model. Due to continuous distribution of the output values, regressor models are used for the predictions. Mean Squared Error (MSE) is used as a common benchmark to evaluate performance of such models. The goal of the model is to minimize the difference between the predicted and the actual (target) value of the output and to average it over the sample size. Also, it helps in changing its parameter to learn from the fundamental hypothesis function that deduces well the untrained data.

As a first step, the presently prepared dataset is used to train the ANN model. The dataset is split into train data and test data. The accuracy of the model is calculated from the R^2 value. After running the simulation in Python 3, it was observed that the R^2 value for train data is 0.914 and 0.965 for the RMR and Ra, respectively. The corresponding R scores for the test data is observed to be 0.907 and 0.856, respectively. The predicted and actual RMR and Ra values of the train data and test data using ANN is shown in Figures 11.4a and b, respectively. It is observed that the train data gives good accuracy as compared to the test data, with a little overlap in test data. As the ANN is known to perform better with a bigger dataset and the currently prepared dataset is based on a full factorial DOE, the training model performance is found to be superior and a lack of fit is noticed between the predicted output and the experimental values in the case of the testing model. The deviation is still higher for the R_a values in comparison to the RMR values.

In the next step, the DTR model is trained using the currently acquired dataset. The accuracy of the model is found to be inferior in comparison to the ANN model.

(a)

(b)

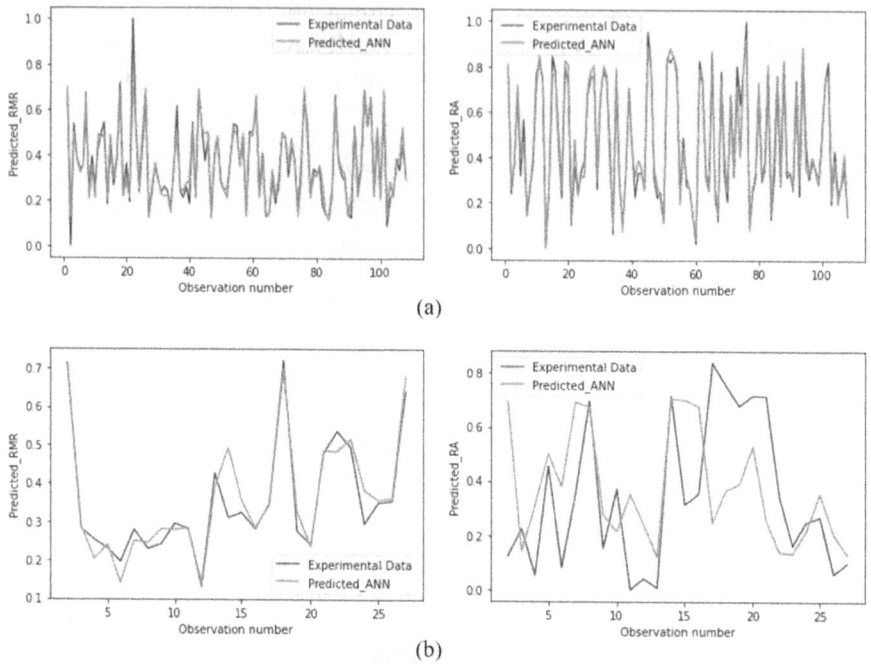

FIGURE 11.4   Predicted RMR and R_a values of (a) trained data and (b) test data using ANN model.

The R^2 value for trained data is found to be 0.912 and 0.928 for the RMR and Ra, which suggests lack of fit of the predicted data and trained data. Similar behaviour is also noticed in case of the test data, where the R^2 value as 0.892 and 0.827 are attained for the RMR and R_a, respectively. Figures 11.5a and b depict the predicted and actual deflection of the train data and the test data, respectively using the DTR model. The results clearly indicate the deficiency in fit of the prediction model with the actual data. Further, the dataset is used to train the RFR model. The R^2 values for the train data and test data are obtained as (0.979, 0.988) and (0.900, 0.879) for the RMR and the R_a, respectively. A similar pattern of overlapping for both the train and the test data is observed in Figures 11.6a and b, those represent the predicted and actual RMR and R_a of train data and the test data, respectively using the RFR model. A better agreement of the predicted and actual RMR and R_a is observed in comparison to the ANN and DTR with superior overlap of the predicted and actual data, both in training as well as testing.

Finally, upon implementing the Light Gradient Boost Regression (LGBR) model, the trained and the test dataset resulted the R^2 value of 0.927 and 0.908, respectively for RMR values. The R scores of 0.973 and 0.892 are attained for R_a, with the training and testing models, respectively. The corresponding distribution pattern is shown in Figure 11.7a and b for RMR and R_a, respectively. Excellent agreement between the predicted and experimental values are noted as compared to the other models. Specifically, the best predictive performance is clearly evidenced from Figure 11.7b

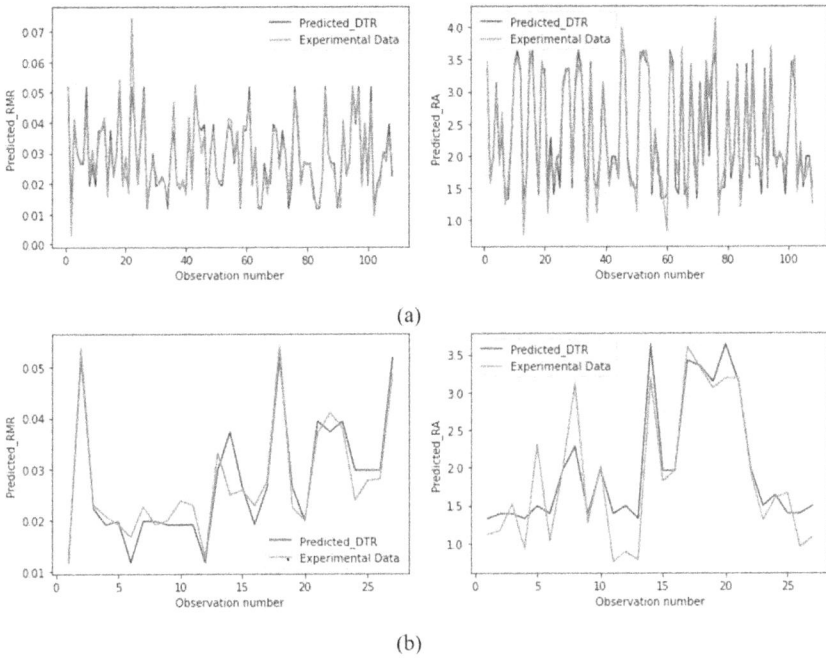

(a)

(b)

FIGURE 11.5   Predicted RMR and R_a values of (a) trained data and (b) test data using DTR model.

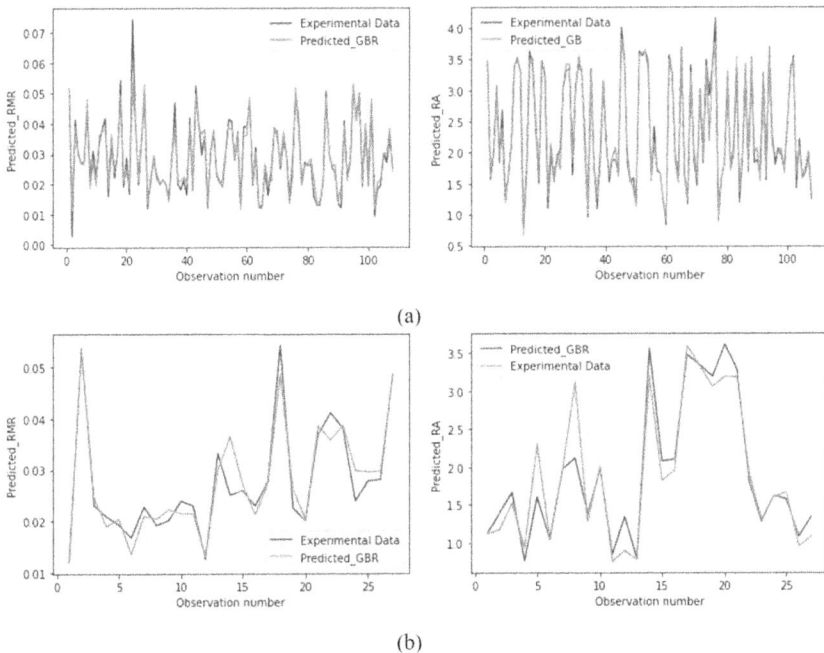

(a)

(b)

FIGURE 11.6   Predicted RMR and R_a values of (a) trained data and (b) test data using RFR model.

(a)

(b)

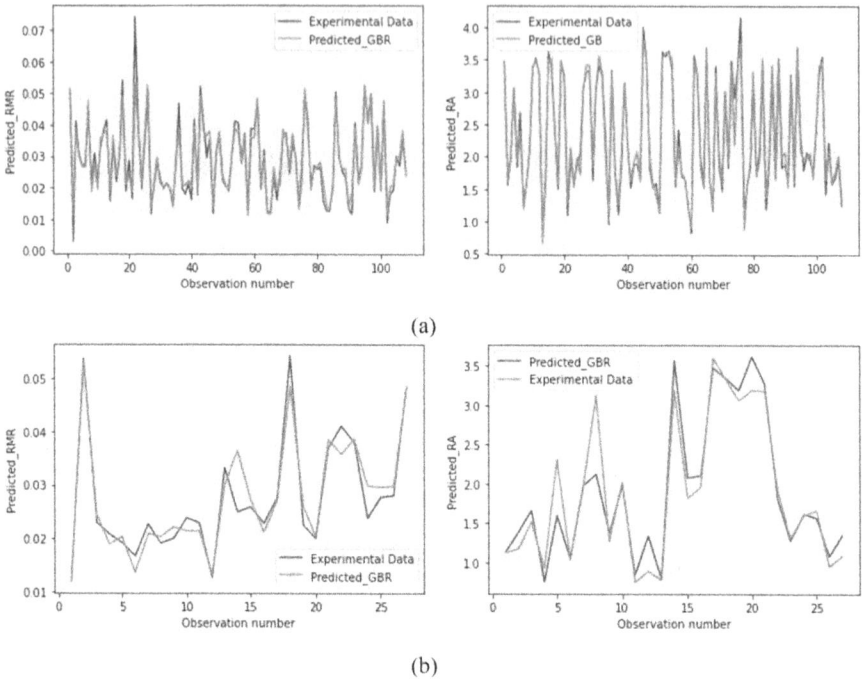

FIGURE 11.7   Predicted RMR and R_a values of (a) trained data and (b) test data using the LGBR model.

TABLE 11.4
Accuracies of the Algorithms Used

Model Deployed	RMR Prediction		Surface Roughness (R_a) Prediction	
	R^2 value (Train)	R^2 value (Test)	R^2 value (Train)	R^2 value (Test)
ANN	0.914	0.907	0.965	0.856
DTR	0.912	0.892	0.928	0.827
RFR	0.979	0.900	0.988	0.879
LGBR	0.927	0.908	0.973	0.892

that depicts the outstanding performance accuracy of the model in terms of the R_a value prediction.

The summary of the present analysis showing the respective accuracies of the algorithms used are tabulated in Table 11.4. It is inferred that the RFR and the LGBR models are providing overall satisfactory results so far as the currently considered dataset is concerned, and the LGBR model can be recommended as the most suited model to predict both the RMR and R_a during dry turning of the graphene reinforced AMMCs. This is also validated by performing confirmation tests randomly

TABLE 11.5

Confirmatory Test Results

Feed (m/min)	RPM (rev/min)	DOC (mm)	wt.% of Graphene	MRR (gm/sec)				
				Actual	GBR	RFR	DTR	ANN
100	3000	0.2	0	0.02658	0.02649	0.02794	0.02835	0.029
150	2000	0.15	0.5	0.03206	0.03003	0.03006	0.02977	0.03477
100	2000	0.2	1	0.02527	0.02607	0.02534	0.02649	0.02638
200	2000	0.15	2	0.0375	0.03731	0.03682	0.03967	0.0358
200	1000	0.2	3	0.04998	0.04981	0.04907	0.05282	0.05182
Feed (m/min)	RPM (rev/min)	DOC (mm)	wt.% of Graphene	R_a (μm)				
				Actual	GBR	RFR	DTR	ANN
100	3000	0.2	0	0.02658	0.02649	0.02794	0.02835	0.02900
150	2000	0.15	0.5	0.03206	0.03003	0.03006	0.02977	0.03477
100	2000	0.2	1	0.02527	0.02607	0.02534	0.02649	0.02638
200	2000	0.15	2	2.33010	0.03731	0.03682	0.03967	0.03580
200	1000	0.2	3	1.65900	1.66788	1.62029	1.52746	1.56663

considering diverse wt.% of graphene reinforcement alongside different combination of controllable parameters. The results are depicted in Table 11.5. Least error in terms of predicting both the RMR and the R_a in comparison to the actual values is clearly noticed for the LGBM model, followed by the RFR model.

11.5 CONCLUSION

In this paper, the modern machine learning techniques are employed to develop the regression models for the prediction of the rate of material removal and average surface roughness values during the dry turning operation of graphene reinforced aluminum metal matrix composites. The metal composites are fabricated by stir casting method with graphene reinforcement at 0, 0.5, 1, 2, and 3 wt.%. Dry turning operations are performed by considering the cutting speed, feed rate, depth of cut, and weight percentage of graphene as input controllable parameters according to full factorial DOE and the desired responses are acquired. With graphene reinforcement, the machinability of the aluminum MMC has been improved in terms of MRR and average surface roughness. Further, the dataset thus prepared are used as the input for various machine learning algorithms. For minimizing the expected error owing to the differences of units of the input parameters, standard scalar has been used to perform scaling of both the training as well as testing data. Moreover, cross validation is executed using GridSearchCV for ensuring best parameters for the models. From the present analysis, it is observed that all the supervised machine learning models performed well in predicting the RMR and R_a, with satisfactory R scores. In the ANN and DTR models the index of R^2 is fewer, which may be attributed to the lack of enough input features that are used to train the models. Some other feature vectors may be added in the future to attain more accuracy. The DTR model is the least accurate in predicting the RMR and R_a, which may be due to less correlation between the parameters. The LGBR model is found to be the best among all the models (followed by the RFR) applied for test data but was a little less accurate than the RFR model in terms of the train data. Based on the confirmatory test results, the LGBR model is found to provide comprehensive accuracy in terms of all the aspects, and therefore, can be endorsed as a suitable model to predict the RMR and R_a during the dry turning of graphene reinforced aluminum metal matrix composites.

REFERENCES

Aggogeri, F., N. Pellegrini, and F. L. Tagliani. 2021. Recent Advances on Machine Learning Applications in Machining Processes. *Applied Sciences* 11 (18): 8764.

Alajmi, M. S., and A. M. Almeshal. 2020. Prediction and Optimization of Surface Roughness in a Turning Process Using the ANFIS-QPSO Method. *Materials (Basel, Switzerland)* 13 (13): 2986.

Alajmi, M. S., and A. M. Almeshal. 2021. Modeling of Cutting Force in the Turning of AISI 4340 Using Gaussian Process Regression Algorithm. *Applied Sciences* 11 (9): 4055.

Behera, R. K., B. P. Samal, S. C. Panigrahi, and K. K. Muduli. 2020. Microstructural and Mechanical Analysis of Sintered Powdered Aluminium Composites. *Advances in Materials Science and Engineering*: e1893475. https://doi.org/10.1155/2020/1893475

Bodunrin, M. O., K. K. Alaneme, and L. H. Chown. 2015. Aluminium Matrix Hybrid Composites: A Review of Reinforcement Philosophies; Mechanical, Corrosion and Tribological Characteristics. *Journal of Materials Research and Technology* 4 (4): 434–45.

Brillinger, M., M. Wuwer, M. A. Hadi, and F. Haas. 2021. Energy Prediction for CNC Machining with Machine Learning. *CIRP Journal of Manufacturing Science and Technology* 35 (November): 715–23.

Caggiano, A. 2018. Tool Wear Prediction in Ti-6Al-4V Machining through Multiple Sensor Monitoring and PCA Features Pattern Recognition. *Sensors* 18 (3): 823.

Cheng, M., L. Jiao, X. Shi, X. Wang, P. Yan, and Y. Li. 2020. An Intelligent Prediction Model of the Tool Wear Based on Machine Learning in Turning High Strength Steel. *Proceedings of the Institution of Mechanical Engineers, Part B: Journal of Engineering Manufacture* 234 (13): 1580–97.

Das, A., S. R. Das, J. P. Panda, et al. 2022. Machine Learning-Based Modeling and Optimization in Hard Turning of AISI D6 Steel with Advanced Altisin-Coated Carbide Inserts to Predict Surface Roughness and Other Machining Characteristics. *Surface Review and Letters* 29 (10): 2250137.

Du, C., C. L. Ho, and J. Kaminski. 2021. Prediction of Product Roughness, Profile, and Roundness Using Machine Learning Techniques for a Hard Turning Process. *Advances in Manufacturing* 9 (2): 206–15.

Dubey, V., A. K. Sharma, H. Kumar, and P. K. Arora. 2022a. "Prediction of Cutting Forces in MQL Turning of AISI 304 Steel Using Machine Learning Algorithm." *Journal of Engineering Research*: 1–13.

Dubey, V., A. K. Sharma, and D. Y. Pimenov. 2022b. Prediction of Surface Roughness Using Machine Learning Approach in MQL Turning of AISI 304 Steel by Varying Nanoparticle Size in the Cutting Fluid. *Lubricants* 10 (5): 81.

Elsheikh, A. H., T. Muthuramalingam, S. Shanmugan, et al. 2021. Fine-Tuned Artificial Intelligence Model Using Pigeon Optimizer for Prediction of Residual Stresses during Turning of Inconel 718. *Journal of Materials Research and Technology* 15 (November): 3622–34.

De Farias, A., S. L. R. de Almeida, S. Delijaicov, V. Seriacopi, and Ed C. Bordinassi. 2020. Simple Machine Learning Allied with Data-Driven Methods for Monitoring Tool Wear in Machining Processes. *The International Journal of Advanced Manufacturing Technology* 109 (9): 2491–2501.

He, K., M. Gao, and Z. Zhao. 2019. Soft Computing Techniques for Surface Roughness Prediction in Hard Turning: A Literature Review. *IEEE Access* 7: 89556–69.

Jurkovic, Z., G. Cukor, M. Brezocnik, and T. Brajkovic. 2018. A Comparison of Machine Learning Methods for Cutting Parameters Prediction in High Speed Turning Process. *Journal of Intelligent Manufacturing* 29 (8): 1683–93.

Khanna, V., V. Kumar, and S. A. Bansal. 2021. Mechanical Properties of Aluminium-Graphene/ Carbon Nanotubes (CNTs) Metal Matrix Composites: Advancement, Opportunities and Perspective. *Materials Research Bulletin* 138 (June): 111224.

Kim, D. H., T. J. Y. Kim, X. Wang, et al. 2018. Smart Machining Process Using Machine Learning: A Review and Perspective on Machining Industry. *International Journal of Precision Engineering and Manufacturing-Green Technology* 5 (4): 555–68.

Lakshmana Kumar, S., V. Jacintha, A. Mahendran, R. M. Bommi, M. Nagaraj, and Umamahesawari Kandasamy. 2022. A Machine Learning Approach to Optimize, Model, and Predict the Machining Factors in Dry Drilling of Nimonic C263. *Advances in Materials Science and Engineering* 2022 (June): e4856089.

Misaka, T., J. Herwan, O. Ryabov, et al. 2020. Prediction of Surface Roughness in CNC Turning by Model-Assisted Response Surface Method. *Precision Engineering* 62 (March): 196–203.

Radhika, N., S. B. Senapathi, R. Subramaniam, R. Subramany, and K. N. Vishnu. 2013. Pattern Recognition Based Surface Roughness Prediction in Turning Hybrid Metal Matrix Composite Using Random Forest Algorithm. *Industrial Lubrication and Tribology* 65 (5): 311–19.

Sekhar, R., T. P. Singh, and P. Shah. 2022. Machine Learning Based Predictive Modeling and Control of Surface Roughness Generation While Machining Micro Boron Carbide and Carbon Nanotube Particle Reinforced Al-Mg Matrix Composites. *Particulate Science and Technology* 40 (3): 355–72.

Shanmugasundar, G., M. Vanitha, R. Čep, V. Kumar, K. Kalita, and M. Ramachandran. 2021. A Comparative Study of Linear, Random Forest and AdaBoost Regressions for Modeling Non-Traditional Machining. *Processes* 9 (11): 2015.

Singh, G., J. P. Appadurai, V. Perumal, et al. 2022. Machine Learning-Based Modelling and Predictive Maintenance of Turning Operation under Cooling/Lubrication for Manufacturing Systems. *Advances in Materials Science and Engineering* 2022 (July): e9289320.

Suthar, J., and K. M. Patel. 2018. Processing Issues, Machining, and Applications of Aluminum Metal Matrix Composites. *Materials and Manufacturing Processes* 33 (5): 499–527.

Tabaszewski, M., P. Twardowski, M. Wiciak-Pikuła, N. Znojkiewicz, A. Felusiak-Czyryca, and J. Czyżycki. 2022. Machine Learning Approaches for Monitoring of Tool Wear during Grey Cast-Iron Turning. *Materials* 15 (12): 4359.

The Royal's Academy of Sciences 2010. Nobel Prize ® and the Nobel Prize ® medal design mark are registered trademarks of the Nobel Foundation Scientific Background on the Nobel Prize in Physics Grapphene compiled by the Class for Physics of the Royal Swedish Academy of Sciences, 50005, October, pp. 0–10

Ulas, M., O. Aydur, T. Gurgenc, and C. Ozel. 2020. Surface Roughness Prediction of Machined Aluminum Alloy with Wire Electrical Discharge Machining by Different Machine Learning Algorithms. *Journal of Materials Research and Technology* 9 (6): 12512–24.

Venkatesan, S., and M. A. Xavior. 2018. Tensile Behavior of Aluminum Alloy (AA7050) Metal Matrix Composite Reinforced with Graphene Fabricated by Stir and Squeeze Cast Processes. *Science and Technology of Materials* 30 (2): 74–85.

Wang, C. C., P. H. Kuo, and G. Y. Chen. 2022. Machine Learning Prediction of Turning Precision Using Optimized XGBoost Model. *Applied Sciences* 12 (15): 7739.

Xiang, S., X. Wang, M. Gupta, K. Wu, X. Hu, and M. Zheng. 2016. Graphene Nanoplatelets Induced Heterogeneous Bimodal Structural Magnesium Matrix Composites with Enhanced Mechanical Properties. *Scientific Reports* 6 (1): 38824.

Zhang, Y., and X. Xu. 2022. Machine Learning Surface Roughnesses in Turning Processes of Brass Metals. *The International Journal of Advanced Manufacturing Technology* 121 (3): 2437–44.

12 Focused Ion Beam Machining as a Technology for Long Term Sustainability

Deepak Patil

National Institute of Technology Tiruchirappalli,
Tiruchirappalli, India

CONTENTS

12.1   INTRODUCTION

Richard Feynman suggested in 1959 that one-day technology would exist that could be used as our eyes and tools in the microscopic world. He anticipated the use of concentrated ion beams to help see and control matter at the smallest scales, in his lecture *"There is plenty of space at the bottom"* (Feynman, 1960), which is often regarded as the genesis of nanotechnology. It took another 40 years before the first widely accessible Focused Ion Beam/Scanning Electron Microscope (FIB/SEM) was available (Young and Moore, 2005). Gallium FIB/SEMs, which combine a scanning electron microscope and a focused ion beam in a single device, have since become essential tools in many institutions and laboratories.

Although tiny silicon devices can be developed using photolithography technology, today's nanoscale devices must also use other materials. One of the most recent processing methods is a focused ion beam (FIB) machining, which first became

DOI: 10.1201/9781003346623-12

widely accessible in the 1990s. Currently, transmission electron microscopy (TEM) sample preparation and semiconductor manufacturing are the two primary applications for FIB processing. It is an essential instrument for design work, failure analysis, lithographic mask repair, device cross-sectioning, maskless implantation, and ion beam-aided etching. Recently, it has been used in a scanning electron microscope (SEM), 3-dimensional patterning with high precision, and improved material removal in microelectronic applications. The fabrication of small structures using FIB technology can create items with micro- or nanoscale features. This in turn sparked study into the fabrication of three-dimensional objects at the micro- and nanoscale. High-resolution imaging has become essential when FIB emerges from failure analysis and replacement mask, particularly towards fabrication. It was essential to use an SEM to examine the ion milled surfaces, hence, FIB and SEM merging has been achieved. Without losing the benefits of each column, milling, deposition, and high-resolution imaging with damage-free imaging are all feasible. To halt the milling operation precisely at the required depth, an ion beam cutting system and an electron beam observing system will be useful (Robert, 1999).

12.2 BASIC FIB EQUIPMENT

Figure 12.1 depicts a standard FIB configuration. Multiple detectors and lens detectors are included in FIB. FIB frequently have manipulators and gas insertion system. Liquid metal ion sources (LMIS), particularly gallium ion sources, are used by the majority of FIB systems. There are also ion generators based on pure gold and

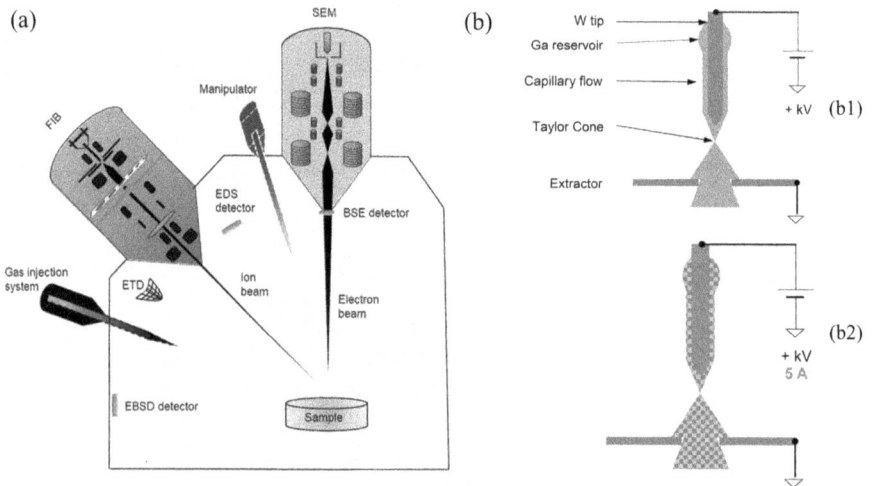

FIGURE 12.1 (a) A typical FIB/SEM system. FIB/SEMs combine a SEM and a FIB into a single apparatus, and they frequently come with a variety of detectors, such as ETD, BSE, EDS, EBSD, and in lens detectors. On FIB/SEMs, manipulators and gas injection devices are frequently found (b) Schematic showing the differences between a Ga-LMIS (b1), in which melting the source metal doesn't require any heating current, and an LMAIS (b2) filled with a eutectic alloy, which needs the tip to be heated with a 5 A current to enable ion emission of both ion species (Annalena, W., 2020; Jacques et al., 2018, Produced with permission).

iridium. In a gallium LMIS, heated gallium metal is put in contact with a tungsten needle. The heated gallium wets the needle, flows to the tip, and is then formed into a Taylor cone (Figure 12.1b) by the conflicting forces of surface tension and the electric field. This cone's point radius is incredibly tiny (about 2 nm). The gallium atoms are ionized and emit electromagnetic fields as a result of the intense electric field at this tiny point. The source ions are then typically focused onto the sample by electrostatic lenses after being accelerated to an energy of 1–50 keV. LMIS creates ion beams with a very tiny energy dispersion and high current density. A contemporary FIB can image a sample with a spot size of a few nanometers or send tens of nanoamperes of current to a sample.

Gallium (Ga) is primarily used as a liquid metal ion source (LMIS) in FIB systems for the reasons listed below (Lucille and Fred, 2005). Swanson (1994) evaluated the efficacy of liquid metal ion sources – Al, Ga, In, Au, and B.

- Ga has a reduced melting point ($T_m = 29.8°C$), allowing for a small design and minimal heating.
- Ga has low fluctuation, which prolongs source life.
- Ga's low surface free energy encourages sticky behavior, which encourages substrate soaking.
- Ga's favorable emission properties allow for high rotational intensities and a narrow energy distribution.
- High luminosity and favorable flow characteristics.
- Its mass is almost perfect because the larger components can be milled thanks to their weight.

Instruments that use plasma beams of noble gas ions, like xenon, have lately become more commonly accessible. To expand LMIS to ion species other than Ga, very early and vigorous efforts have been made. For instance, switching N to P dopants released from the same liquid metal alloy ion source would enable semiconductor dopant implantations (Bischoff et al., 2016). The need to choose appropriate ion species for doping or sample processing, friendly methods for better control of ion-induced population, selective insertion of ion species, or the need for particle deposition, are all made possible by the FIB system. To achieve this, an ongoing endeavor has been made to broaden the LMIS to other metals and alloys over the years, opening the door to the potential of up to 46 different ionic species.

12.2.1 Principle

FIB systems operate in a similar fashion to a scanning electron microscope (SEM) except, rather than a beam of electrons and as the name implies, FIB systems use a finely focused beam of ions that can be operated at low beam currents for imaging or at high beam currents for site-specific sputtering or milling. The gallium (Ga^+) main ion beam strikes the sample surface as shown in the diagram on the right, and a tiny quantity of material sputters out, leaving the surface as secondary ions (i+ or i) or neutral atoms. Secondary electrons (e) are also produced by the mainstream. The information from the sputtered ions or secondary electrons is gathered as the main beam rasters on the sample surface to create an image.

Modern FIB systems can readily achieve 5 nanometer imaging resolution at low main beam currents because very little material is sputtered (imaging resolution with Ga ions is restricted to 5 nm by sputtering and detector efficiency) (Koch et al., 1999). Sputtering can eliminate a significant amount of material at greater main currents, enabling accurate milling of the specimen down to a submicrometer or even a nanoscale.

The interplay between the energetic ions and the solid material's center or electrons when the ion stream bombards its surface can result in a variety of physical and chemical events. As a consequence, a succession of events may take place, as illustrated in Figure 12.2, including backscattering, implantation, excitation, amorphization, sample heating, and chemical reactions, in addition to the solid atoms being sputtered out. As a result, mastering the fundamentals of ion-solid interactions could significantly improve one's capacity to create superior nanostructured surfaces using FIB nanofabrication technology. From the viewpoint of ion energy, the occurrence of the aforementioned ion-solid interactions can be seen as a group of elastic collision and inelastic collision events. As a consequence, the following intriguing features can be seen in Figure 12.2a. When an elastic impact takes place between an incident ion and a target atom within a collision cascade area, incident ions are reflected to create backscattered ions. The energy of the incoming ions is transmitted in part to the colliding atoms to produce secondary electrons (SEs) and X-rays, while the target atoms are energized and ionized to produce visible light, ultraviolet light, and infrared light when an inelastic impact takes place. Additionally, the solid substance surface can release secondary ions when the incident ions attack it (see Figure 12.2b). These phenomena are capable of carrying out tasks like spectrum analysis and ion beam photography. If the kinetic energy delivered by the ions is large enough to exceed the surface binding energy, the collision process can cause the target atoms of the solid lattice to become off-site atoms, which may pass through the lattice gap and be ejected as sputtered particles from the surface of the material (see Figure 12.2c). In order to create tension or strain, the target material is subjected to ion implantation and material amorphization, which may also cause the target material's structure to distort (see Figure 12.2d). Ion beam deposition can be carried out by adding external circumstances, such as gas molecules when the energy supplied by the ions is less than the surface binding energy (see Figure 12.2e). Another scenario is that the incident ions may impart energy and velocity to the atoms affixed to the solid material's surface, causing the atoms to penetrate the solid material further. The primary processes of ion implantation involve these two physical collisions, both of which alter the material's surface characteristics (see Figure 12.2f). Elastic impact or inelastic collision can also start chemical reactions in the solids, and the latter is particularly important in such chemical reactions if the irradiated substance is a compound. A low-energy electron deluge cannon can be used to provide charge neutralization if the material is non-conductive. In this way, even highly insulating materials can be imaged and milled without a conducting surface layer, as would be necessary in an SEM, by imaging with positive secondary ions using the positive primary ion stream.

Up until lately, the semiconductor sector was where FIB was most widely used. Applications like defect analysis, circuit alteration, photomask repair, and the production of site-specific TEM samples for integrated circuits have all become standard practices. Modern FIB systems have high resolution imaging capabilities, and when

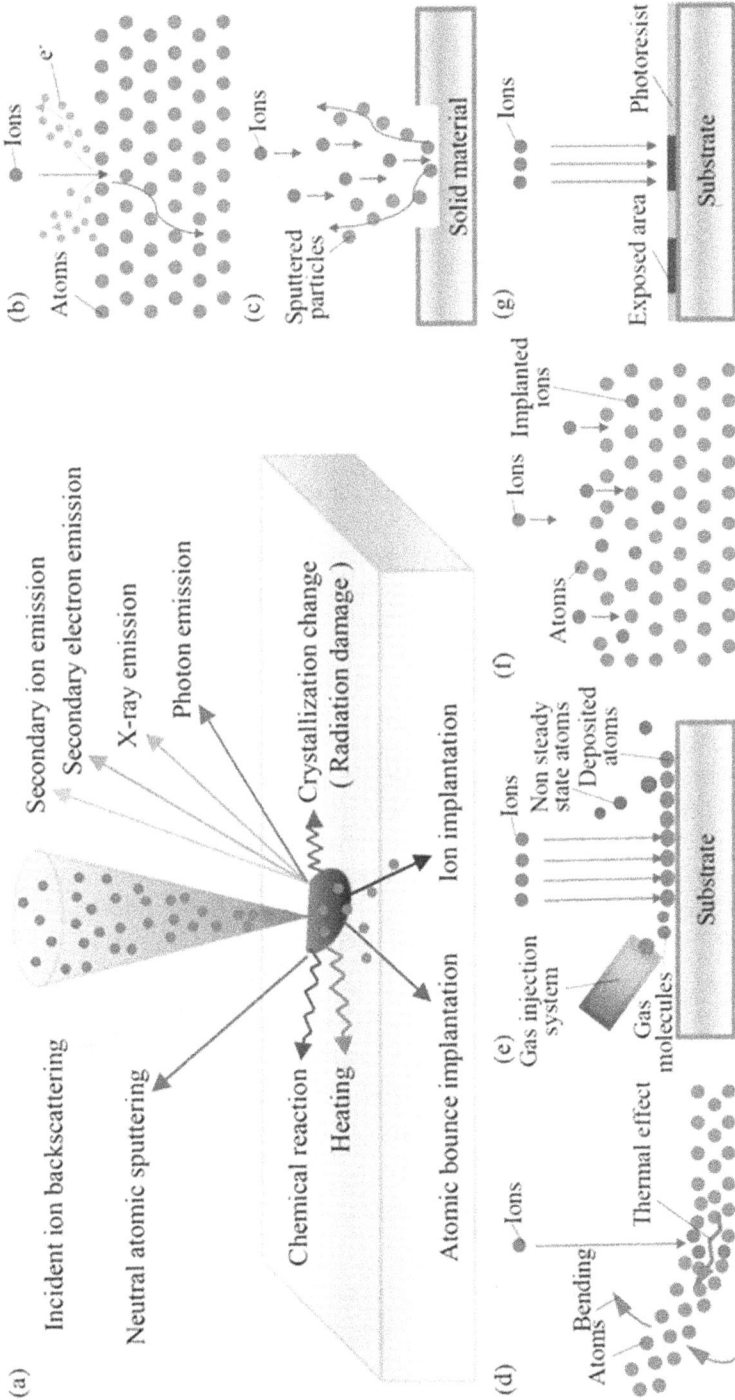

FIGURE 12.2 The fundamental method by which the ion stream interacts with a solid substance is shown in (a). Schematic representations of the tiny interaction process between ions and atoms or molecules in lithography, etching, irradiation, deposition, implantation, and (b) imaging are shown in figures (b) through (f). (Reproduced with permission, Ping et al., 2021).

combined with in situ sectioning, they have often rendered unnecessary the need for SEM examination of FIB sectioned materials (Aravindan et al., 2010). SEM imaging is still needed for the greatest resolution imaging and to avoid harm to delicate samples. However, the advantages of both can be used when SEM and FIB columns are attached to the same room.

12.2.2 Process Parameters in FIB Process

Modern FIB machines are computer-controlled, and grinding is done with exact pixel-by-pixel movements. The procedure, also referred to as a computer scan, is roughly depicted in Figure 12.3.

The FIB tracks a string of neighboring pixels that indicate the channel's contour in order to mill a submicron channel. Ion flow, fluence, dose, beam current, and current density are the main terms used to characterize the properties of an ion beam. The temporal rate of energy movement is known as flux. The quantity of particles moving through a specific region in a particular amount of time is known as flux. The ions/cm^2 measure flux. The word "dose" refers generally to the amount of radiation or particle absorption by a material. The same measure of fluence applies to dose. The quantity of ions that travel through a specified area before striking an object within that area is known as fluence. The range of the beam current is a few pA to several nA. The energy level or quantity of ions in a given region at any given moment is measured by the current density. It uses the columns/cm^2 measure (Orloff et al., 2003). The relationship between beam current, dwell point spacing, and dwell duration determines how many ions are given to the substrate per unit area (ion dose). For a specific substrate substance, the ion dose can be converted into the profundity of sputtering. The Ga+ beam current, stay duration, and pixel size together decide the pixel dosage or the number of ions per unit area. The tiniest element of a digital picture is typically a pixel. The Ga$^+$ beam current, machine zone size, and machining duration are used to calculate the overall Ga$^+$ dosage. Scanning is the technique used to trace tightly separated parallel lines for sputtering. The TRIM modeling of the relationship between the sputtering yield and incident angle at a 90 keV ion energy

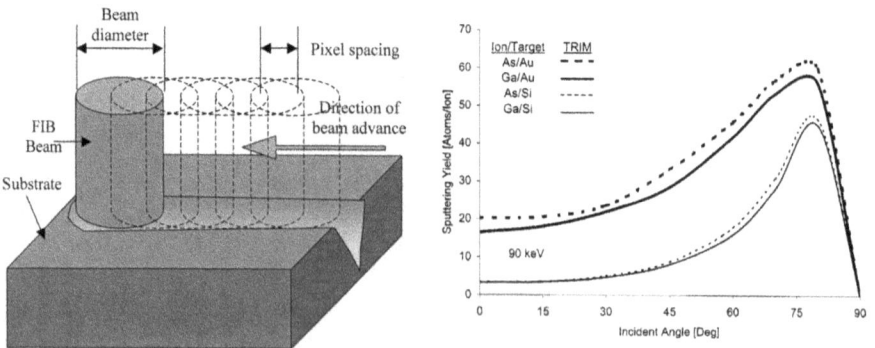

FIGURE 12.3 (a) Schematic of FIB milling, (b) Angular dependence of sputtering yields of 90 keV As and Ga ions on Au and Si targets (Reproduced with permission, Ampere et al., 2004).

in the sputtering of Au and Si surfaces is shown in Figure 12.3b. In the modeling, a 90 kV electric potential is used to single charge and energize the As and Ga atoms. According to Figure 12.3b, for all instances taken into consideration, the sputtering yield increases as the incidence angle increases until it hits its maximum near 80°, at which point it quickly declines to zero as the incident angle approaches 90°. Figure 12.3b also demonstrates that while the equivalent sputtering yield for the Au substrate rises less, that of the Si substrate increases roughly ten times from the normal incidence to the angle at its apex.

12.3   MULTI-PROCESS CAPABILITY OF FIB

12.3.1   FIB ETCHING

FIB is intrinsically harmful to the specimen, unlike an electron microscope. The sample will discharge atoms from the surface when the high-energy gallium ions hit it. Additionally, the upper few nanometers of the surface will be injected with gallium atoms, causing the surface to become porous. The FIB is used as an instrument for micro- and nano-machining materials at the micro- and nanoscale because of its sputtering capacity. Although FIB nano milling is still in its infancy, it has grown into a sizable subfield of its own. Typically, 2.5–6 nm is the lowest beam size for imaging. The interactions between the milling sample and the overall beam size determine the size of the tiniest milled features, which are slightly bigger (10–15 nm).

12.3.2   GAS ASSISTED FIB DEPOSITION

Ion beam induced deposition can also be used with a FIB to place substance. When a gas, such as tungsten hexacarbonyl ($W(CO)_6$), is introduced to the vacuum container and permitted to chemisorb onto the sample, FIB-assisted chemical vapor deposition takes place. The precursor vapor will split into volatile and non-volatile components as the laser scans an area, The non-volatile component, like tungsten, deposits on the surface. This is advantageous because the metal can be used as a protective layer to shield the sample beneath from the beam's harmful sputtering. Metal bands of any length, from nanometers to hundreds of micrometers, can be deposited using tungsten. A local deposition is also a possibility for other elements like platinum, cobalt, carbon, gold, etc. (Gorji et al., 2020). Table 12.1 summarizes the different precursor gases used for depositing different materials.

TABLE 12.1
FIB Deposition Precursor Gases

S. No.	Element to Be Deposited	Precursor Gas
1	Carbon	Phenanthrene
2	Platinum	Tri methyl cyclo penta dienyl platinum
3	Tungsten	WCO_6 Tungsten hexa carbonyl
4	SiO_2	O_2 and tetra methoxy silane or O_2 and tetraetoxysilane
5	Aluminium	Trimethylamine alane

Source: Aravindan et al., 2010

In the semiconductor business, FIB is frequently used to patch or alter an existing semiconductor device. The gallium beam, for instance, could be employed in an integrated circuit to remove undesirable electrical connections and/or to apply conductive substance to create connections. In patterned doping of semiconductors, the high degree of surface contact is taken advantage of. FIB is also employed for insertion without a covering.

12.3.3 FOR TRANSFER OF SENSITIVE SAMPLES

For a minimal introduction of stress and bending to transmission electron microscopy (TEM) samples (lamellae, thin films, and other mechanically and beam sensitive samples), when transferring inside a focused ion beam (FIB), flexible metallic nanowires can be attached to a typically rigid micromanipulator. This method's primary benefits include a substantial reduction in sample preparation time (rapid nanowire welding and cutting at low beam current), as well as a decrease in stress-induced bending, Pt contamination, and ion beam damage.

12.3.4 FIB TOMOGRAPHY

A potent instrument for site-specific 3D imaging of sub-micron characteristics in a material is now the focused ion beam. In this method of FIB tomography, the sample is successively milled with an ion beam pointed perpendicular to the specimen while an electron beam images the freshly exposed surface. Larger size nanostructures can be characterized using this so-called "slice and view" method in a variety of SEM imaging modalities, such as secondary electron, backscattered electron, and energy dispersive X-ray measurement. Since the object is being progressively milled away after each picture is taken, the procedure is destructive. After recording the picture stack and eliminating artifacts, the gathered sequence of images is then converted into a 3D volume. Ion mill curtaining, where mill patterns create broad aperiodic bands in each picture, is the main artifact that impairs FIB tomography. Destriping methods can be used to eliminate the ion mill curtaining. FIB tomography can be carried out on materials and living samples, and at both ambient and cryogenic conditions (Hartnell et al., 2016).

12.4 APPLICATIONS OF FIB SYSTEM

Figure 12.4 shows how FIB milling has been applied to micro-nano fabrications. The development of tools and dies for micromachining, microforming, useful surfaces through surface structuring, the creation of microsurgical instruments, etc., all make use of high ion momentum beams. In comparison to a cutting tool without any structures, the microtextured cemented carbide tool greatly lowers the cutting force by 28% when used for traditional machining of aluminum alloy. By decreasing the chip-tool contact length, the microtextures serve as a lubricant reservoir and lessen friction at chip-tool surfaces. FIB-SEM have been widely used for recording live 3D imaging of substantial living materials in recent times. The 3D images of live cells are shown in Figure 12.5. Table 12.2 summarizes the how different functionality of FIB can be used for different applications.

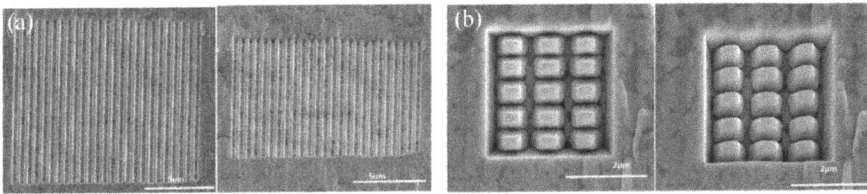

FIGURE 12.4   250 nm pitch and depth nanogrooves (FIB process parameters: 30 keV, 0.5 nA, rest time: 10 s, incidence angle: 0°). (A) Top perspective with a 20° angle. (FIB process parameters: 30 keV, 0.5 nA, rest time: 10 s, angle of incidence: 0°) Nanopillar array with 500 nm side length and 500 nm depth (b) upper image with a 20-degree tilt   (Produced with permission, from Goswami et al., 2018)

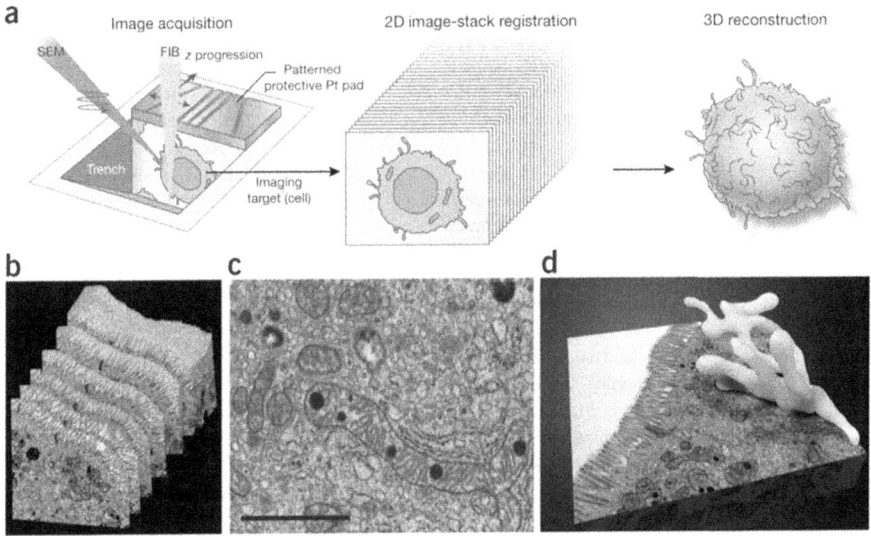

FIGURE 12.5   FIB-SEM 3D imaging of substantial living materials. (a) Large biological samples are inserted into the FIB-SEM apparatus after being traditionally fixed (by aldehydes) or cryogenically fixed (by high-pressure refrigeration), stained with heavy metals, resin embedded, and mounted. Here, selected sample regions are "trenched" to expose the region of interest. The region of interest is then imaged using a SEM (blue beam) and FIB (yellow beam) to create a 2D image stack after a repeated cycle of resin milling. The material to be scanned is protected by a patterned platinum (Pt) pad that enables automated beam adjustment and slice-thickness control. The 3D structure of interest is then revealed by mathematically converting the 2D picture stack to a 3D volume, aligning, and segmenting it. (b–d) A rodent intestinal sample was used as a typical case of 3D tissue imaging. An image stack (b), a chosen slice through the stack (c), and a segmented depiction of a mitochondrion with many branches that is present in the scanned volume are all displayed. (d) 1 millimeter scale marker (Reprinted with consent from Elsevier from Hartnell, L.M. et al., 2016).

TABLE 12.2

An Overview of 3D Nanostructure and Device Manufacturing Using FIB Nanofabrication

FIB Assisted Processes	Application Field	Typical Components
Etching	Semiconductor devices and large-scale integrated circuits, sample detection, high-power solid-state laser, renewable energy, nano-optics, single molecule detection	Optical mask, transistor, chip, TEM samples, crack detection, micro-sensor, nanosensors, etc.
	Sample detection, integrated circuits, MEMS, nano-optics, optoelectronic devices, molecular devices, Raman spectroscopy, electrochemical, biomedical	Gold electrodes, third-generation DNA sequencers, advanced extreme ultraviolet lithography mask repair, etc.
	Electrical, optical, optoelectronic platform, magnetic, medical diagnosis, MEMS, solar–thermal energy harvesting system	Resonator, Fresnel zone plates, multifeatured shaped 3-dimensional nanostructures, silicon nano-line structure, ordered nano-textured surface, surface optical functional texture of crystalline silicon cells, etc.
Deposition	Integrated circuit, nano-optics, nanoplasmonics, electronics, nanosensing, microelectrode system, microfluidics, semiconductors and nanosuperconductor, complex surface structures	Optical masks, X-ray masks, and optical phase masks, ohmic contact layer, connection line, diffraction grating, nanochannels and hollow nanowires, micro-hemispheres, parabolic and sinusoidal microstructures, etc.
Implantation	Nanoelectronics, nanophotonics, plasmonics, sensing, biomedical, circuitry, semiconductor, surface modification, nano-cutting, device performance regulation, size analysis, and new device production	optical waveguide, nanochannels, freestanding bridges and cantilevers, resonators, suspended beams, silicon-based nanostructures, micro-pyramid anti-reflection layer, micro-pillar array, InGaZnO thin-film transistors, thin circuitry, etc.

Source: Goswami et al., 2018; Aravindan et al., 2010; Koch et al., 1999.

The most typical function is milling. These tools have a wide range of applications, including the production of atomic force microscope tips (Annalena, 2020), the creation of nanopillars for mechanical testing, the cross-sectioning and failure analysis of integrated circuits, and the fabrication of nanostructures for devices and sensors.

12.5   SUMMARY

Future materials can now be researched and developed using a variety of ion species and focused ion beams (FIBs). From site-specific cross-sectioning and 3D models to TEM lamellae preparation, nanofabrication, and defect engineering, there is an FIB tool for every application field. The choice of the best ion species/system for the job is aided by the basic physics of ion solid interactions as well as the system's available tools (such as gas injection systems, manipulators, and detectors). Virtually any material type can be milled and viewed in a fluid manner, down to submicron to nanometer scale. Due to its higher precision and reduced lateral scatter, FIB is preferable to other micromachining methods when compared.

REFERENCES

Feynman, P. 1960. There's plenty of room at the bottom, *Engineering and Science* 23: 22–36.

Young, R. J., Moore, M. V. 2005. Dual-Beam (FIB-SEM) Systems. In *Introduction to Focused Ion Beams*, ed. L. A. Giannuzzi, F. A. Stevie, 247–268, Springer, Boston, MA.

Robert, K. 1999. Dual column (FIB-SEM) wafer application, *Micron* 30: 221–226.

Annalena, W. 2020. Focused ion beams: An overview of the technology and its capabilities. *Wiley Analytical Science Magazine*, https://analyticalscience.wiley.com/do/10.1002/was.00070009

Swanson, L. W. 1994. Use of liquid metal ion source for focused ion beam applications. *Applied Surface Science* 76–77: 80–88.

Lucille, A. G., Fred, A. S. 2005. *Introduction to focused ion beams, instrumentation, theory, techniques and practice*, 1–347, Springer Inc, Boston, USA.

Orloff, J., Utlaut, M., Swanson, L. 2003. *High resolution focused ion beams: FIB and Its applications*, 1–303, Springer Press. ISBN 978-0-306-47350-0.

Bischoff, L., Mazarov, P., Bruchhaus, L., Gierak, J. 2016. Liquid metal alloy ion sources—An alternative for focussed ion beam technology, *Applied Physics. Review* 3: 021101.

Ping, L., Siyu, C., Houfu, D., Zhengmei, Y., Zhiquan, C., Yasi, W., Yiqin, C., Wenqiang, P., Wubin, S., Huigao, D. 2021. Recent advances in focused ion beam nanofabrication for nanostructures and devices: fundamentals and applications, *Nanoscale*, 13: 1529–1565.

Koch, J., Grun, K., Ruff, M., Wernhardt, R., Wieck, A.D. 1999. Creation of nanoelectronic devices by focused ion beam implantation. *IECON '99 Proceedings. The 25th Annual Conference of the IEEE*. Vol. 1. pp. 35–39. doi:10.1109/IECON.1999.822165. ISBN 0-7803-5735-3

Aravindan, S., Rao, P.V., Yoshino, M. 2010. *Focused ion beam machining, In Introduction to Micromachining*, 2nd ed. V.K. Jain, Narosa Publication.

Gorji, S., Kashiwar, A., Mantha, L. S., Kruk, R., Witte, R., Marek, P., Hahn, H., Kübel, C., Scherer, T. 2020. Nanowire facilitated transfer of sensitive TEM samples in a FIB, *Ultramicroscopy* 219: 113075.

Goswami A., Umashankar, R., Gupta, A.K., Aravindan, S., Rao, P.V., 2018. Development of a microstructured surface using the FIB, *Journal of Micromanufacturing* 1: 53–61.

Hartnell, L. M., Earl, L. A., Bliss, D., Moran, A., Subramaniamet, S. 2016. Imaging cellular architecture with 3D SEM, *Organizational Cell Biology*, 27: 44–50.

Ampere, A. T., Ivan A. I., Jong S. P., 2004. Milling of submicron channels on gold layer using double charged arsenic ion beam, *Journal of Vacuum Science & Technology B*, 22: 82.

Jacques, G., Paul, M., Lars, B., Ralf, J., Lothar, B. 2018. Review of electrohydrodynamical ion sources and their applications to focused ion beam technology, *Journal of Vacuum Science & Technology B*, 36: 06J101.

Index

Pages in *italics* refer to figures and pages in **bold** refer to tables.

Index

Pages in *italics* refer to figures and pages in **bold** refer to tables.

For Product Safety Concerns and Information please contact our EU
representative GPSR@taylorandfrancis.com
Taylor & Francis Verlag GmbH, Kaufingerstraße 24, 80331 München, Germany